COLUMBIA REVIEW

HIGH-YIELD PHYSICS

COLUMBIA REVIEW

HIGH-YIELD PHYSICS

Stephen D. Bresnick, M.D.
President and Director
Columbia Review, Inc.
San Francisco, California

Williams & Wilkins
A WAVERLY COMPANY

BALTIMORE • PHILADELPHIA • LONDON • PARIS • BANGKOK
BUENOS AIRES • HONG KONG • MUNICH • SYDNEY • TOKYO • WROCLAW
1996

Editor: Elizabeth A. Nieginski
Managing Editor: Alethea H. Elkins
Production Coordinator: Danielle Santucci
Designer: Ashley Pound Design
Typesetter: Maryland Composition Co., Inc.
Printer: Port City Press
Binder: Port City Press

351 West Camden Street
Baltimore, Maryland 21201-2436 USA

Rose Tree Corporate Center
1400 North Providence Road
Building II, Suite 5025
Media, Pennsylvania 19063-2043 USA

Printed in the United States of America

Library of Congress Cataloging in Publication Data

The Publishers have made every effort to trace the copyright holders for borrowed material. If they have inadvertently overlooked any, they will be pleased to make the necessary arrangements at the first opportunity.

96 97 98 99
1 2 3 4 5 6 7 8 9 10

RECYCLED

Contents

SECTION I

Physics of Solids and Fluids

SECTION II Physics of Heat, Electricity, and Magnetism

Physics of Waves, Sound, Light, and Nuclear Structure

SECTION III

Review Questions

Preface

High-Yield General Physics is an **easy-to-read, efficient,** and **high-quality** review book for first-year, non–calculus-based, college-level physics. The book focuses on a conceptual review of core physics topics and covers an amazing amount of material for its size. For mastery of review material, over 250 review questions with solutions are provided. The book is designed for all college students or others wishing to understand and review the major concepts of physics. Students who are pre-health, science, or non-science majors will benefit from this book.

High-Yield Physics is one of four books in the *High-Yield College Science Review* series by Williams & Wilkins. The series also contains *High-Yield General Chemistry, High-Yield Organic Chemistry,* and *High-Yield Biology.* This series has been designed to make these four important college sciences easier to understand and master. All the High-Yield books contain a great science review, many examples and sample problems, and several hundred practice questions with answers and explanations.

The author of this series, Dr. Stephen Bresnick, is an expert in helping students understand, review, and retain basic college science material. Dr. Bresnick understands that many students work their way through college courses without really comprehending the material they are supposed to be learning. He has designed these four books to help students **understand science better** and **improve their course grades.** In addition, the series has been designed to help students prepare for **postgraduate and preprofessional tests,** such as the GRE, MCAT, DAT, PCAT, VET, OAT, and other tests. Dr. Bresnick is a physician and educator who both teaches and writes science review materials for college students. He is currently Director of **Columbia Review,** a national test-preparation company specializing in science and English review for premedical school students.

Organization

There are three sections in this book. Each section corresponds to the specific topics that most college students study in physics courses. The topic review emphasizes conceptual learning and provides numerous sample problems and examples. At the end of the review chapters, there are ten sets of review questions with solutions so that students can assess their comprehension.

Acknowledgments

The author wishes to thank Dr. William Bresnick and Dr. Mark Sornson for their contributions. In addition, many thanks to the staff of Williams & Wilkins for their dedication in creating a great high-yield review book for physics. I especially wish to thank Lee Elkins, Danielle Santucci, Elizabeth Nieginski, Jane Velker, Tim Satterfield, and Kevin Thibodeau for their expertise and assistance with this important project.

SECTION I

Physics of Solids and Fluids

calars

Scalars and Vectors

The physics review notes in this book provide a collection of the relationships and formulas that you need to know. Conceptual review is emphasized. It is important to understand these formulas and concepts; do not just memorize the material. **Success in physics requires conceptual understanding and the ability to apply knowledge to solve problems.**

I. Scalars

A scalar quantity measures only **magnitude.** Scalars are specified with a single number that has particular units. Scalars may be positive or negative and may be added together using simple arithmetic.

Common examples of scalar quantities include **mass, distance, volume, time, work, and power**.

II. Vectors

A vector quantity measures both **magnitude and direction.** These quantities are usually represented graphically by arrows, the heads of which point in the direction of the quantity and the length of which represents the magnitude of the quantity. Vector magnitude is always a positive number.

Common examples of vector quantities include **displacement, force, velocity, acceleration, momentum, gravity, E-fields, and B-fields.**

Two vectors are equal if they have the same magnitude and direction. They must point in the same direction and have equal arrow lengths.

The easiest way to add vectors is with the **graphic method.** Two vectors are added graphically by drawing the first vector and placing the tail of the second vector at the head of the first, keeping the same magnitude and direction. The sum of the two vectors is a vector that begins at the tail of the first and extends to the head of the second.

Figure 1-1 shows the addition of vector a and vector b to give the resultant vector c. Note that to generate vector c, the tail of vector b was placed at the head of vector a, maintaining magnitude and direction.

The easiest way to solve a problem involving the subtraction of vectors is to change the direction of the vector being subtracted by 180° and then add the two vectors: $a - b = a + (-b)$.

A second way to add vectors is to use the **component method,** which involves resolving each vector into its scalar components, adding the x and y components individually, and then using the Pythagorean theorem to find the magnitude of the resultant.

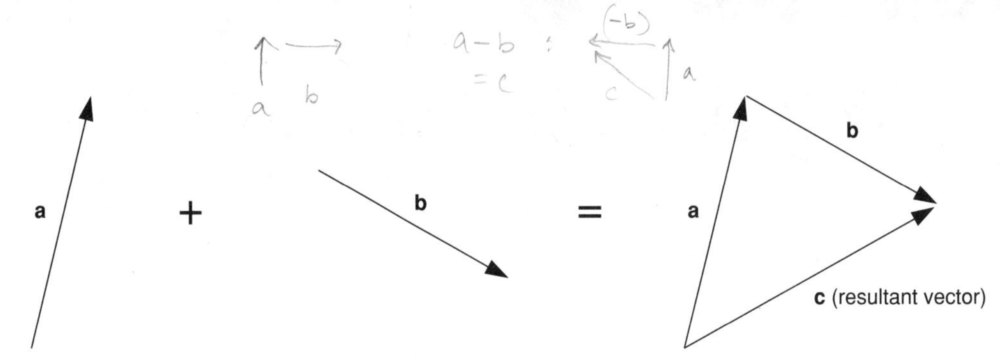

Figure 1-1. Graphic addition of vectors.

This technique is also useful when adding more than two vectors. To find the resultant force of several vectors, add the x components of the force vectors and add the y components of the force vectors. Then, use the formula:

$$\mathbf{F_R} = (\mathbf{F_x}^2 + \mathbf{F_y}^2)^{1/2}$$

Once you find the magnitude of the resultant, you can find the direction of the resultant vector by using the following formula, in which θ is the number of degrees from the x-axis:

$$\tan \theta = (y \text{ component})/(x \text{ component})$$

III. Additional Points to Remember

Some important relationships to commit to memory are those between the sides of right triangles (Table 1-1). Even if you forget the value of one of the sine or cosine values of the angles given in Table 1-1, you can always derive these values if you know the basic triangle relationships. Also included is the **3,4,5 triangle,** the sine and cosine values of which are often used in problems (Figure 1-3).

EXAMPLE 1-1

FIND THE MAGNITUDE and direction of the sum of vectors A and B.

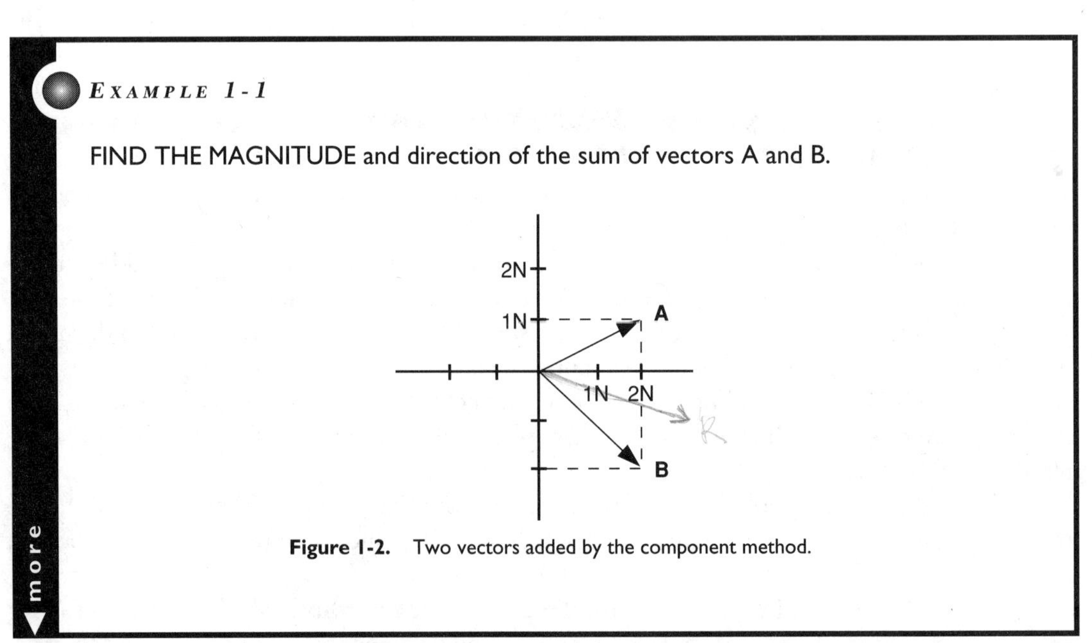

Figure 1-2. Two vectors added by the component method.

Ex. 1-1 cont.

SOLUTION

In Figure 1-2, note that vector A has an *x* component of 2 N and a *y* component of 1 N. Vector B has an *x* component of 2 N and a *y* component of −2 N. The sum of the *x* components is 4 N and the sum of the *y* components is −1 N. Now use the Pythagorean theorem to find the resultant.

$$F = \{(4\text{ N})^2 + (-1\text{ N})^2\}^{1/2} = 4.1\text{ N}$$

$$\tan\theta = \text{y component/x component} = -1\text{ N}/4\text{ N} = -0.250$$

$$\theta = \arctan(0.250) = -14^\circ \text{ (Arctan values would be given on a test)}$$

Thus, the resultant vector has a magnitude of 4.1 N and is directed 14° below the *x*-axis.

TABLE 1-1. Sine and Cosine Values of Right Triangle Relationships

	Relationship	0°	30°	45°	60°	90°
Sin	opp/hypot	0	0.5 or 1/2	0.71 or $\sqrt{2/2}$	0.87 or $\sqrt{3/2}$	1
Cos	adj/hypot	1	0.87 or $\sqrt{3/2}$	0.71 or $\sqrt{2/2}$	0.5 or 1/2	0

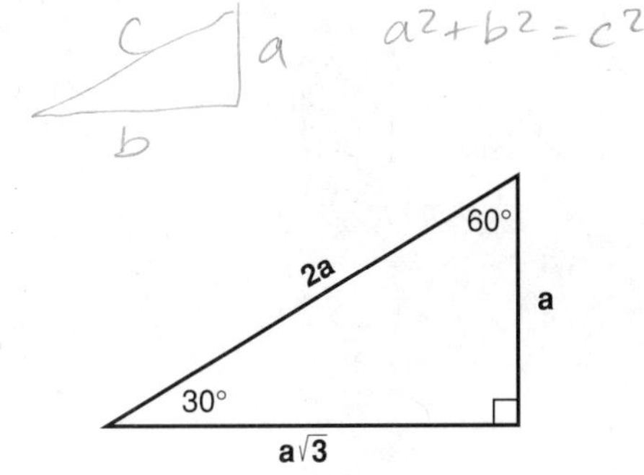

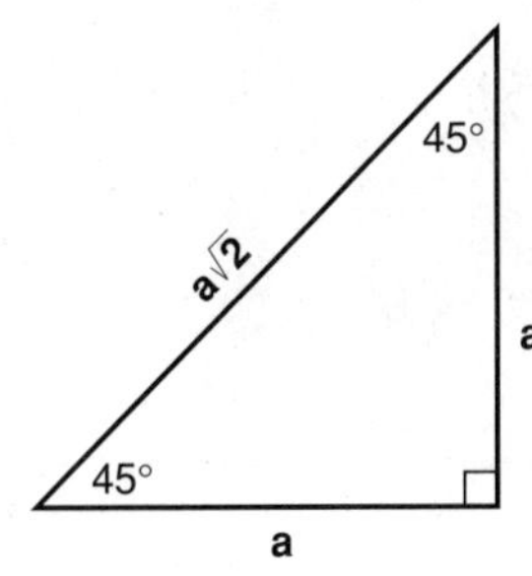

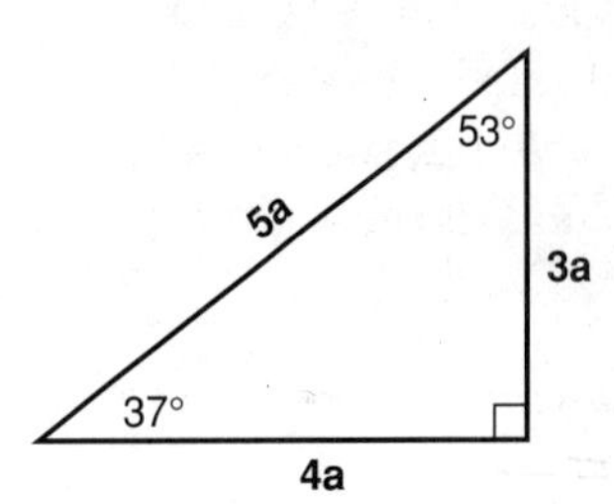

Figure 1-3. Important triangle relationships.

Statics 2

I. Static Systems Without Rotation

Static systems are not moving. The sum of the forces acting on the system **balance** such that the sum of the x components equals zero and the sum of the y components equals zero. Thus, in static systems:

$$\boldsymbol{\Sigma F_x = 0 \text{ and } \Sigma F_y = 0}$$

Static systems require that forces be in equilibrium. The key to solving these problems is to pick a point at which all the forces on the system act.

II. Static Systems with Rotation—Torque

Some static systems, when disturbed, want to rotate. Rotation occurs if forces act in such a way as to generate a net turning effect, or torque. This action may occur if a pulley or suspending rope is removed from a system with an axis of rotation.

Torque is a force multiplied by the perpendicular distance from the place where the force is applied to the axis of rotation. Torque is considered positive if it involves counterclockwise rotation and negative if rotation is clockwise. Taking the sine of the angle gives its component perpendicular to the lever arm. To find the torque of a body with the axis of rotation P, use the following:

Torque = (lever arm)(Force)(sin of angle between body and force)

or

$$\textbf{Torque} = \mathbf{rF}\sin\theta$$

Figure 2-5 shows a nut being turned with a wrench. Assume the wrench is being pulled downward—a great example of applying torque. The axis of rotation, or point P, is at the nut. The radius is the distance from the nut to the hand. The force vector is in the direction of the forearm (down). The angle (θ) is the angle between r and F.

The lever arm is the distance from the place where the force is applied to the axis of rotation. Note that taking the sine of the angle described gives the perpendicular distance from the force to the axis of rotation.

Another way to calculate torque, if you do not want to memorize the equation for torque using the sine function, involves using basic trigonometry and triangle relationships to determine

EXAMPLE 2-1

A BLOCK HANGS at rest from wall surfaces A and B (Figure 2-1). The block is suspended by three weightless ropes. The tension in the rope attaching to wall A is 15 N and the angles the ropes make to the horizontal are 30° for A and 60° for B. Find the tension in the rope attaching to wall B and the weight of the block.

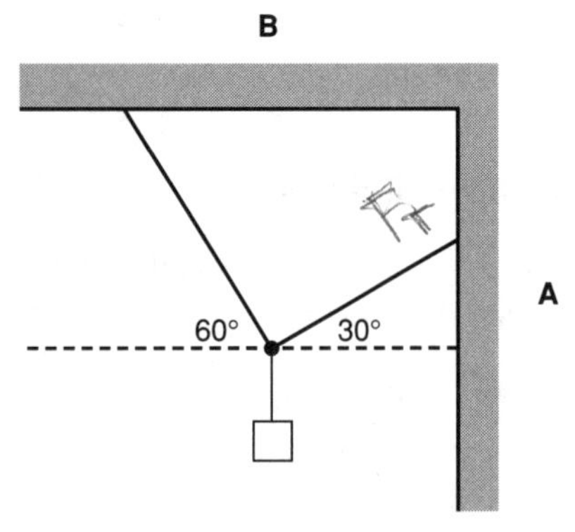

Figure 2-1. A block at rest, supported by three massless ropes.

SOLUTION

Start by finding the point at which all the forces act. Note that all the ropes act at the intersection point; the rope suspending the block is counteracted by the ropes attached to the walls pulling up on the block. The system does not move because the forces in the *x* direction acting at this intersection point all sum to zero. Similarly, the forces acting in the *y* direction all sum to zero at this point. Draw a vector diagram showing the net forces (Figure 2-2).

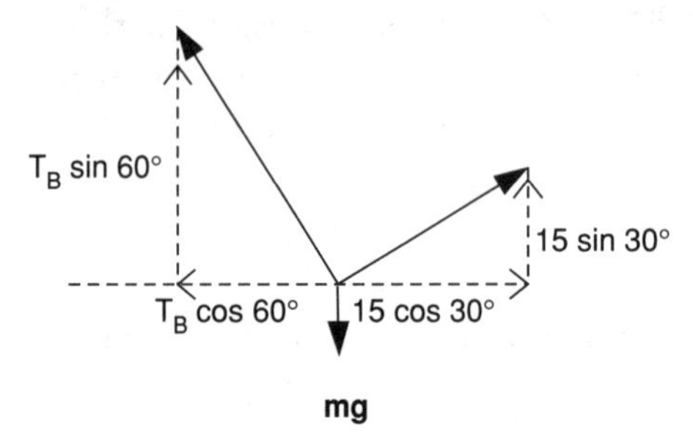

Figure 2-2. Vector diagram of net forces for Example 2-1.

You must sum all the *x* forces, which are vectors that have direction as well as magnitude. The magnitude of the forces is solved by components. If the vector points in the $+x$ axis, give the magnitude a positive sign. If the vector points in the $-x$ axis, give the magnitude a negative sign. The same principle applies for the $+y$ and $-y$ axis for up and down directions.

Therefore,

$\sum F_x = 0.$ $\quad$ $\sum F_x = 15(\cos 30°) + (-T_B \cos 60°) = 0$

$\sum F_x = 15(0.87) + -T_B(0.5) = 0$

so, $\quad T_B \approx 26$ N

Ex. 2-1 cont.

Because both of the ropes attached to the walls are pulling up on the intersection point, they will be added to the force associated with the rope pulling down on this point because of the weight of the block.

Now,

$$\sum F_y = 0$$
$$\sum F_y = 15(\sin 30°) + T_B(\sin 60°) - W = 0$$
$$\sum F_y = 15(0.5) + 26(0.87) - W = 0$$
$$W = 7.5 + 22.6$$
$$W \approx 30 \text{ N}$$

EXAMPLE 2-2

IN FIGURE 2-3, find the magnitude of the tension in ropes *x* and *y*.

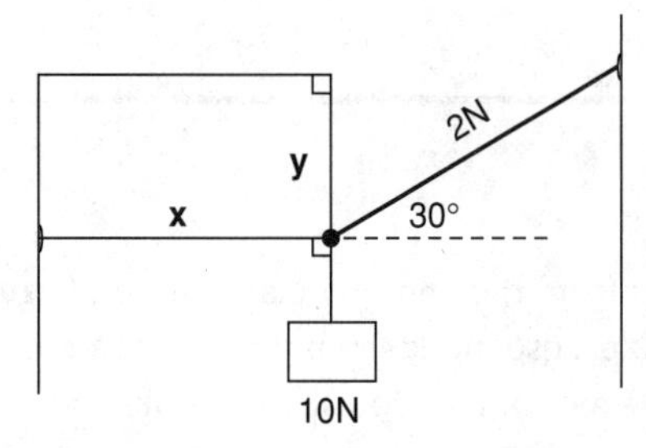

Figure 2-3. Ropes supporting a block.

SOLUTION

Write equations showing the balance of the sum of the *x* and *y* forces. Remember that the angles used to calculate the components of forces are those angles that the ropes make with the horizontal. Start by drawing a vector diagram (Figure 2-4).

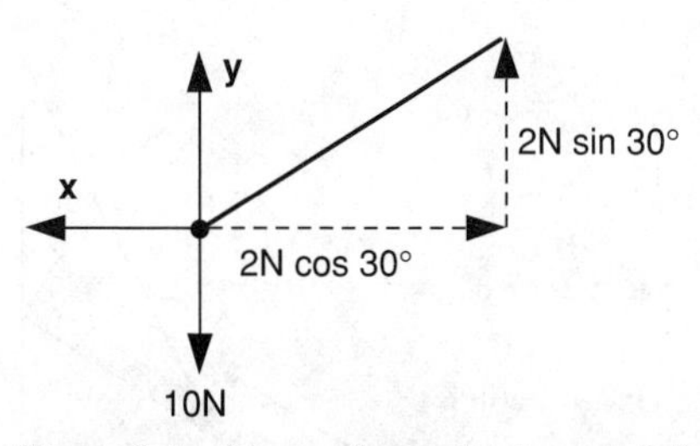

Figure 2-4. Vector diagram for Example 2-2.

$$F_x = 2\text{ N}(\cos 30°) + (-x)(\cos 0)$$
$$F_x = 2\text{ N}(0.87) + (-x)(1) = 0$$
so, $x = 1.74$ N

$$F_y = (y)(\sin 90°) + (2\text{ N})(\sin 30°) + (-10\text{ N})(\sin 90°) = 0$$
$$F_y = (y)(1) + (2\text{ N})(0.5) + (-10\text{ N})(1) = 0$$
so, $y = 9$ N

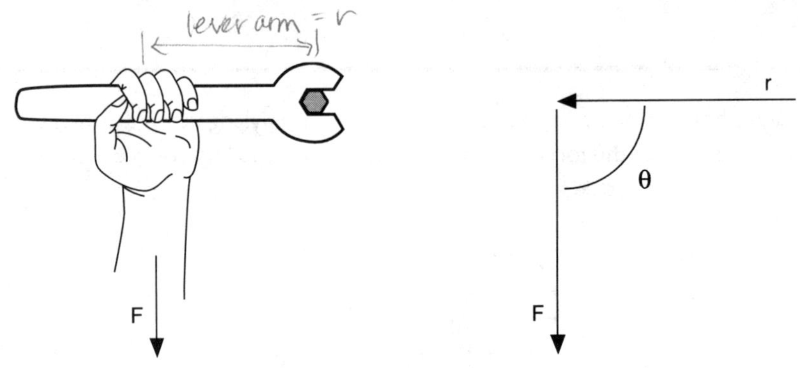

Figure 2-5. Concept of torque. At left, torque is being applied to turn a nut with a wrench. At right, diagram showing F, r, and θ.

$\tau = F \times d$ (lever arm) $F \times r$

the perpendicular distance from the axis of rotation to the applied force. Torque is simply the product of the applied force and the magnitude of this perpendicular distance.

For a potentially rotating system to be a static system, the torques acting on the system must sum to zero. This concept is useful for solving two types of problems, namely, statics problems with a rotational component and center of gravity problems.

EXAMPLE 2-3

A 100-N WEIGHT hangs from the end of a massless rod (Figure 2-6). The rod makes a 60° angle with the wall. A cable, suspended from the wall at a 30° angle, attaches to the rod and aids in its suspension. The axis of rotation is where the rod contacts the wall. The length of the rod is *l* units long.

1. Find the torque produced by the weight
2. Find the torque produced by the cable
3. Find the tension in the cable supporting the rod

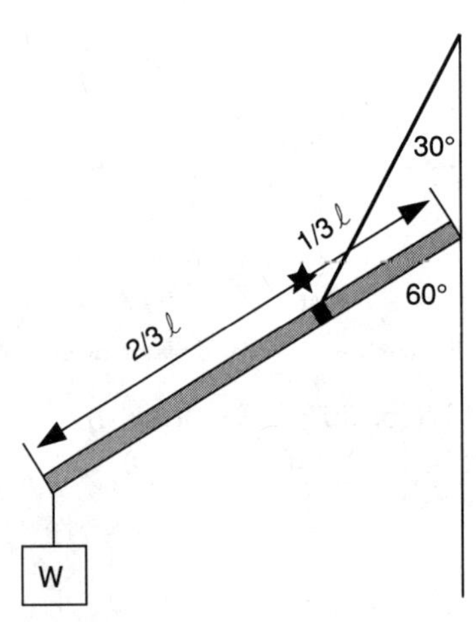

Figure 2-6. Torque problem in Example 2-3. Note that the block is suspended from a massless rod, which is in turn hinged to a supporting wall. A rope also connects the rod to a higher position on the wall.

more

Ex. 2-3 cont.

SOLUTION

This problem is somewhat difficult because the angles can be confusing. When calculating torque, picture in your mind what each applied force does in rotating the body. In this case, an applied weight makes the rod want to rotate counterclockwise about the axis of rotation (positive torque). The supporting cable applies a force (tension) pulling the rod upward and making it want to rotate clockwise about the axis of rotation (negative torque).

1. The torque related to the weight is found by using the equation provided previously in this chapter. Determining the magnitude of angle θ is difficult. Draw a line from the end of the rod back to the wall, making it perpendicular to the wall. Note that the angle between the rod and this line is 30° and the angle between the weight and the line is 90°. Therefore, the angle between the rod and the weight is 120°.

$$\text{Torque} = rF\sin\theta = (l)(100\text{N})(\sin 120°) = 87l$$

2. The key to solving this part is to know which angles are important. If you are to use the torque equation described previously, you need to use the angle between the F and r terms. The force that causes rotation is supplied by the cable. The "r" refers to the line between the axis of rotation and the point at which the cable attaches to the wall. The angle formed is 30°. Thus, the torque from the cable is $-(1/3\ l)(T)(\sin 30°) = -lT/6$

Notice that you are to leave the answer in terms of T. Also note that the rotation is clockwise, giving a negative torque.

3. Take the sum of the torques and set them equal to zero. Using the information from questions 1 and 2, the torque from the weight is $87l$ and the torque from the cable
is $-lT/6$:

$$87l + (-lT/6) = 0$$

$$T = 522\text{ N, answer}$$

CONCEPTUAL QUESTION

A plumber is trying to turn a bolt with a wrench, but it seems that the bolt is rusted in place. He tries to design a way to increase the torque applied to the bolt with his wrench. Which, if either, would you advise him to do: place a long piece of pipe on the handle of the wrench to lengthen the effective handle length, or tie a long rope to the end of the handle to give a longer line of force in which to help turn the wrench?

SOLUTION

The torque applied to the bolt depends on the force applied and the lever arm over which this force is applied. The bolt acts as a pivot point. Figure 2-5 showed the bolt/wrench system and the forces applied. The length of the wrench is the lever arm if the wrench is pulled at a right angle to the line of action of the applied force. Thus, torque can be increased by lengthening the wrench handle. Tying a rope to the end of the wrench to help turn the bolt may not increase the force applied to the lever arm.

III. Center of Gravity and Center of Mass

It is possible to use the concept of torques to find the center of gravity. The torque about any point produced by the weight of an object is equal to that from a concentrated object of the same weight placed at a point called the **center of gravity.** By definition, the center of gravity of an object is that point at which you may consider the total force of gravity to act.

The **center of mass** of an object is that point at which the total mass is concentrated. A suspended object always hangs so that its center of gravity is directly below the point of suspension. In this position, the torque resulting from the weight about that point is zero. It is intuitive that the center of gravity and the center of mass are usually located at the same point. The only instance in which they would not be equal is if the value of gravity varied over the volume of the object.

To calculate the center of mass, break up the object into small sections and take the center of mass of each section. The easiest way to find the center of mass of an object made up of two separate weights is to use the following formulas:

$$X = (x_1w_1 + x_2w_2)/w_{tot} \text{ or } X = (x_1m_1 + x_2m_2)/m_{tot}$$

in which X is the location of the center of gravity based on your chosen *x*-axis coordinate system, x_1 is the distance of the first weight from the chosen *x*-axis coordinate system, x_2 is the distance of the second weight from the chosen *x*-axis coordinate system, w_{tot} is the total weight of the object in question ($w_1 + w_2$), and m_{tot} is the total mass of the object in question ($m_1 + m_2$).

A similar formula can be set up for the *y*-axis. Just substitute *y*'s for the *x*'s in the preceding equation.

EXAMPLE 2-4

Consider the two-dimensional shape in Figure 2-7 as having two halves. Each half has a point of mass concentration at the center of mass. Point A is located 2 cm above and 2 cm to the right of the lower left-hand corner of the shape. Point B is located 3 cm above and 8 cm to the right of the lower left-hand corner of the shape. Point A has a weight of 20 N, whereas point B has a weight of 30 N. Find the center of mass of the entire shape.

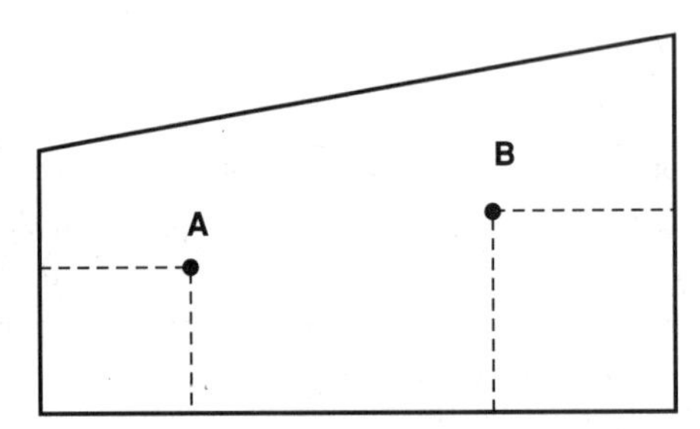

Figure 2-7. Center of mass example.

SOLUTION

Use a coordinate system for the *x*-axis and *y*-axis, and suppose that the lower left-hand corner of the shape has the coordinates (0,0). Solve for the center of mass for the *x*-axis and *y*-axis independently.

$X = (x_1w_1 + x_2w_2)/w_{tot}$ $X = \{(2\text{ cm})(20\text{ N}) + (8\text{ cm})(30\text{ N})\}/(20\text{ N} + 30\text{ N}) = 5.6\text{ cm}$

$Y = (y_1w_1 + y_2w_2)/w_{tot}$ $Y = \{(2\text{ cm})(20\text{ N}) + (3\text{ cm})(30\text{ N})\}/(20\text{ N} + 30\text{ N}) = 2.6\text{ cm}$

Thus, the center of mass of this object is (5.6, 2.6) on the chosen coordinate system.

Kinematics

kinema

I. Basic Concepts

Kinematics is the study of the motion of objects and their path of travel. Some definitions are critical to your understanding of this subject.

The **displacement of an object** is the shortest straight line connecting its starting point and its ending point, no matter what its path. Displacement is a vector quantity, and shows both magnitude and direction. Do not confuse displacement with distance, which is a scalar quantity and describes the total length of travel, not the shortest path.

The **velocity** of a point is its net displacement divided by the time it takes the displacement to occur. It may also be written as:

$$\mathbf{v} = (\mathbf{x_f} - \mathbf{x_i})/\Delta \mathbf{t}$$

in which x_f and x_i are final and initial positions, respectively.

The **average speed** of an object is its total distance traveled divided by the total elapsed time.

Constant speed is a constant change of distance covered over time. To say that an object traveled 20 m in 4 seconds at a uniform rate without mention of direction is to describe constant speed. **Constant velocity** infers travel at a constant rate in a constant direction, with zero acceleration.

Average acceleration is the change in velocity divided by the time required for the change to occur. **Instantaneous acceleration** is the average acceleration over a very short time interval.

$$\mathbf{a} = \Delta \mathbf{v}/\Delta \mathbf{t}$$

Acceleration does not infer going faster or slower. It merely states that velocity is changing over time. An object may accelerate if its rate of travel is constant, yet its direction of velocity is changing. A prime example of this principle is discussed in Chapter 7 (Circular Motion).

II. Graphic Relations

To grasp the true meaning of the terms used in this chapter, study the following graphs. Clear understanding of the graphic representation of kinematic concepts is important.

EXAMPLE 3-1

CONSIDER THE PATH followed by a boomerang that returns to its thrower. If the boomerang travels 10 m before returning to its thrower within 10 seconds, find the average speed and average velocity of the boomerang.

SOLUTION

The average speed is the total distance divided by elapsed time.

$$\text{Total distance} = 20 \text{ m}/10 \text{ sec} = 2 \text{ m/sec}$$

The average velocity is zero because there is no net displacement.

The first graph (Figure 3-1) plots displacement over time. Each portion of this graph represents the following:

Section a: A period of rest, or no motion. Note that the displacement does not change over a period of time.
Section b: A forward velocity which appears constant as the slope of the line of displacement/time appears constant.
Section c: A second period of rest.
Section d: Acceleration. Note that the change in displacement/time appears to be an exponential function, and that the slope of lines tangential to points on the curve are increasing.

Figure 3-2 shows velocity versus time. Each section of this graph represents the following:

Section a: Constant velocity (no acceleration).
Section b: Constant and negative acceleration (deceleration) or change in velocity divided by change in time.
Section c: Rest, because velocity is zero at this point.
Section d: Increasing acceleration or acceleration increasing over time.

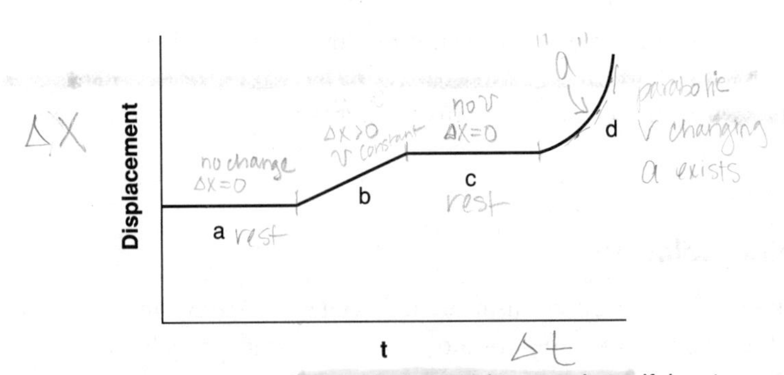

Figure 3-1. Displacement versus time for an object. Slope = displacement/time = velocity. If slope is constant, velocity is constant. If slope is not constant, velocity is not constant and there is acceleration.

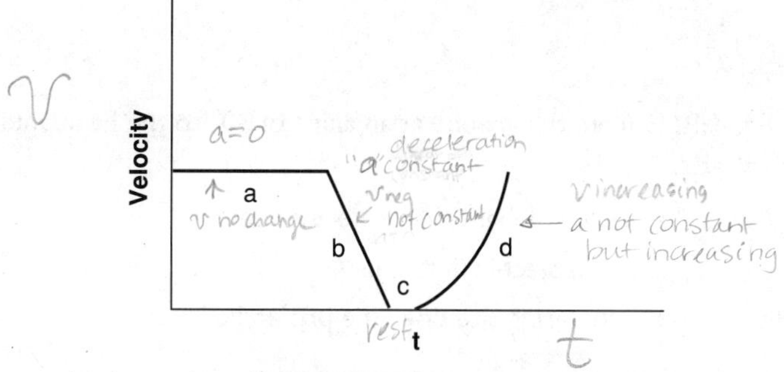

Figure 3-2. Velocity versus time for an object. Slope = velocity/time = acceleration. If slope is constant, then acceleration is constant. If slope is not constant, acceleration is not constant.

Sample Problems

EXAMPLE 3-2

A CAR TRAVELING at a constant 100 m/sec for 10 seconds suddenly accelerates at 10 m/sec^2 for 10 seconds. How far does the car travel in the 20-second time period?

SOLUTION

For the **first 10 seconds:** $x = vt = (100 \text{ m/sec})(10 \text{ sec}) = 1000 \text{ m}$

For the **second 10 seconds,** you must take acceleration into account:

$$x = v_o t + 1/2\, at^2 \quad \text{(distance during the acceleration)}$$
$$x = (100 \text{ m/sec})(10 \text{ sec}) + (1/2)(10 \text{ m/sec}^2)(10 \text{ sec})^2$$
$$x = 1500 \text{ m} \quad \text{(distance for acceleration period)}$$

Total distance covered = 1000 m + 1500 m = 2500 m

EXAMPLE 3-3

A BULLET FIRED from a gun decelerates at 200 m/sec^2 in its first 1000 m of travel. If it covers this distance in 1.5 seconds and has a velocity of 500 m/sec at 1000 m, find the velocity at which the bullet left the gun.

SOLUTION

$$v_f^2 = v_o^2 - 2ax$$
$$(500 \text{ m/sec})^2 = v_o^2 - (2)(200)(1000)$$
$$250{,}000 = v_o^2 - 400{,}000$$
$$650{,}000 = v_o^2$$
$$v_o \approx 800 \text{ m/sec}$$

EXAMPLE 3-4

A PROJECTILE IS FIRED from the ground at an angle of 30° to the horizontal with an initial velocity of 10 m/sec.

1. What is the greatest height the projectile reaches?
2. How long is the projectile in the air?
3. How far from the firing site does the projectile land?

SOLUTION

Draw a vector diagram (Figure 3-3). You must find the *x* and *y* components of the initial velocity. Use the 30-60-90 triangle rule.

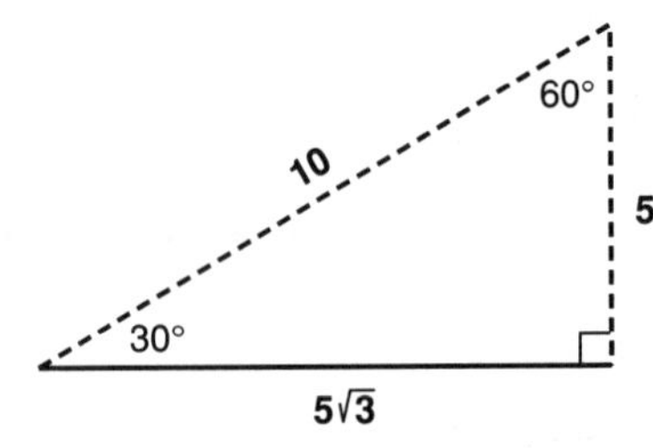

Figure 3-3. Vector diagram for Example 3-4.

1. To find the greatest height the projectile reaches, use the kinematic equation that has y as a variable.

$$v_{fy} = 0 \quad \text{(object is at rest at its highest point)}$$
$$v_{oy} = 10 \sin 30° = 5 \text{ m/s } (y \text{ component of initial velocity})$$
$$a_y = -10 \text{ m/sec}^2$$
$$v_{fy}^2 = v_{oy}^2 - 2ay$$
$$0^2 = (5 \text{ m/sec})^2 - (2)(10 \text{ m/sec}^2)(y)$$
$$y = 1.25 \text{ m}$$

2. Now find the time the projectile is in the air.

$$v_f = v_0 + at$$
$$0 = \text{m/sec} - (10 \text{ m/sec}^2)(t)$$
$$5 = 10\,t$$
$$t = 0.5 \text{ sec}$$

3. How far does the object travel in the horizontal direction? You know that the object reaches its highest point at 0.5 sec, so it again strikes the ground in another 0.5 sec. **The path is symmetric,** so the time it requires an object to reach its highest point when fired from the ground is equal to the time it requires to fall from the high point back to the ground. For this problem, total travel time is 1.0 sec. The horizontal distance traveled is the product of the *x* component of the initial velocity and the time in the air. Note that the *x* component of the initial velocity is constant throughout the travel of this projectile because there is no acceleration in the *x*-direction acting on the object.

$$x = (v_o)_x(t)$$
$$x = (5\sqrt{3} \text{ m/s})(1 \text{ sec}) = 5\sqrt{3} \text{ m}$$

EXAMPLE 3-5

A STONE IS THROWN from the roof of a 10-m high building at an angle of 45° above the horizontal (Figure 3-4). If it strikes the ground 10 seconds later at a distance of 100 m from the base of the building, find:

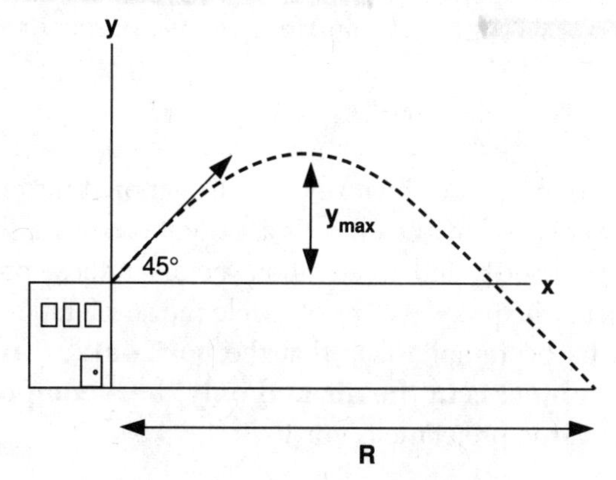

Figure 3-4. Diagram for Example 3-5.

1. The initial velocity of the stone
2. The maximum height reached above the roof

SOLUTION

1. Find the x component of the initial velocity. Because the stone traveled 100 m in 10 seconds in the x-direction, $v_x = 10$ m/sec.

$$v_x = v_o \cos 45°$$
$$10 \text{ m/sec} = (v_o)(0.71)$$
$$v_o = 14.1 \text{ m/sec}$$

2. Now find the vertical component of the initial velocity.
Recall that the final velocity in the y-direction at the highest point (y_{max}) is zero.

$$v_f^2 = v_o^2 + 2ay$$

Knowing that

$$v_{oy} = v_o \sin 45° = (14.1 \text{ m})(\sin 45°) = 10 \text{ m/sec.}$$
$$0 = (10 \text{ m/sec})^2 + (2)(-10 \text{ m/sec}^2)(y)$$
$$y = 5 \text{ m}$$

III. Kinematic Equations

Several equations are useful for straight-line motion with constant acceleration.

$$\mathbf{x = 1/2(v_f + v_o)(t)} \qquad (1)$$

in which v_0 and v_f are initial and final velocities, respectively. Equation 1 makes sense because the average velocity should approximate the average between initial and final velocities.

$$\mathbf{v_f^2 - v_o^2 = 2ax} \qquad (2)$$

$v^2 = v_o^2 + 2ax$

in which x is the distance traveled. Equation 2 helps you find the final velocity if you know the initial velocity, the acceleration of the object, and the distance traveled.

$$x = v_o t + 1/2\ at^2 \quad (3)$$

One of the most useful for problem solving, equation 3 allows you to relate distance traveled with acceleration and time traveled. Remember to place a negative sign in front of the $1/2at^2$ term if the acceleration is gravity (g). The value for g is -9.8 m/sec^2 or approximately -10 m/sec^2.

$$v_f = v_o + at \quad (4)$$

This formula is the definition of final velocity. Equation 4 makes sense because the initial velocity of the object is enhanced by the object's acceleration over a short unit of time.

The directions of the velocity and acceleration vectors in these problems differ. The velocity vector is tangent to the path the projectile travels while the acceleration vector is *down* (gravity).

Note: It cannot be overemphasized that the **horizontal travel of the projectile depends on the time the object is in the air and only the *x* component of its velocity.** The only acceleration acting on the projectile is gravity.

Conceptual Questions

1. As a ball rolls down a hill (Figure 3-5), how do its speed and acceleration change?

   Solution: The speed of the ball increases as it rolls down the hill. Acceleration, however, depends on how steep the hill is. Acceleration is the greatest at the top of the hill because the hill is steepest. As the ball rolls down the hill, the hill is less steep and the acceleration decreases. Thus, acceleration can decrease while speed increases.

2. You drop a small marble and a large bowling ball at the same time from the same height. Ignoring air resistance, why do both of these objects fall together with equal accelerations?

   Solution: Ignoring air resistance, all free-falling objects fall with the same acceleration (-9.8 m/sec^2), given Newton's second law: $F = ma$ or $a = F/m$. Free-falling objects are subject to the force of gravity, which gives an object's weight. Thus, $a = W/m$. Because weight and mass are proportional, the relative increase in both of these values for large, heavy objects tends to cancel out in the $a = W/m$ expression. For example, a bowling ball that has 50 times the weight of a marble also has 50 times the mass. Because acceleration is the quotient of force (weight) and mass of falling objects, the effect of the greater weight and mass of the bowling ball compared to the marble cancels out, and both the bowling ball and marble fall with the same acceleration.

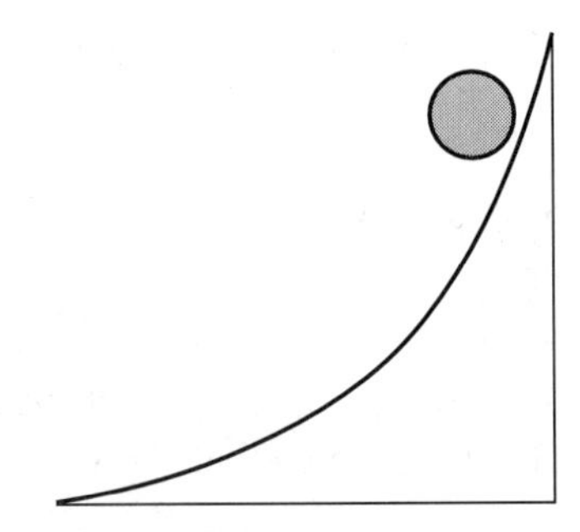

Figure 3-5. Diagram for conceptual question 1.

Dynamics

4

......dynami

Dynamics is the study of how things move. This chapter offers a review of the basic concepts of dynamics, including force, Newton's laws of motion, and friction.

I. Force and Weight

Force is a mass times an acceleration. Force is what causes the acceleration of an object, commonly thought of as a "push" or a "pull." Newton's second law states:

$$\mathbf{F = ma}$$

The terms of force in the SI system are a combination of base units of mass, length, and time, called the Newton (N). A Newton is force that imparts an acceleration of 1 meter per second to a 1-kg mass.

$$\mathbf{N = (kgm)/s^2}$$

Weight is the downward force an object experiences on or near the surface of the earth. Weight is the product of the mass of the object and the acceleration of gravity.

$$\mathbf{W = mg}$$

Remember that the weight of an object is exclusively the gravitational force exerted by the earth on an object. On different planets, an object will maintain the same mass, but may have different weights, because the value of g differs on other planets.

II. Newton's Laws of Motion

Newton's first law says that an isolated object at rest will remain at rest if no net force acts on it. If an isolated object is in motion, it will continue moving along a straight line at constant speed if no net force acts on it.

The first law emphasizes no net force. Several forces acting on an object that balance one another to produce zero net force are viewed the same as if no force is acting on the object at all.

Newton's second law says that if the sum of all forces on an object is not zero, then the object will be accelerated. The acceleration produced depends on the sum of the forces and on the mass of the object. Simply stated: F = ma.

Newton's third law says that for every force, there is an equal and opposite reaction force. If object A exerts a force on object B, then object B exerts an equal and opposite force on object A.

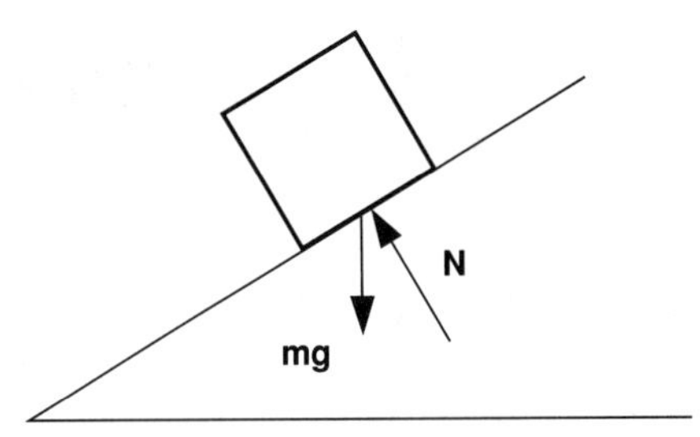

Figure 4-1. Normal force (N) is a reaction force directed perpendicular to the plane on which an object rests. Note the difference between the normal force and the weight (mg).

This law addresses the concept of normal force or N. A normal force is a reaction force that is perpendicular to the plane on which an object rests (Figure 4-1). If a block rests on a flat tabletop, normal force is directed straight up. If a block sits on a ramp, the normal force is directed perpendicular to the plane of the ramp.

III. Friction

As a block resting on a flat surface is subjected to force, the object usually slides along that surface. This movement occurs only if the force applied is greater than the friction force between the block and the surface. **Friction force** opposes the tendency to move when one object slides over another.

The friction force (f) is the coefficient of friction (μ) times the normal force.

$$f = \mu N$$

The coefficient of friction is a **property of the surfaces** that contact each other. The coefficient of static friction (μ_s) is involved if two or more objects have friction forces between them and no motion occurs between the surfaces. If motion occurs between the surfaces, it is known as the coefficient of kinetic friction (μ_k). **For a given pair of surfaces, the coefficient of kinetic friction is always less than the coefficient of static friction.**

Each of these friction coefficients is associated with a force. The force of **kinetic friction** is parallel to the surface of contact of the two objects, although the direction of the kinetic friction force is always opposite the velocity of the object. The **force of static friction** keeps an object from moving as a result of an applied force. The direction of this force is opposite the direction of the applied force. Figure 4-2 shows the direction of each of these friction forces.

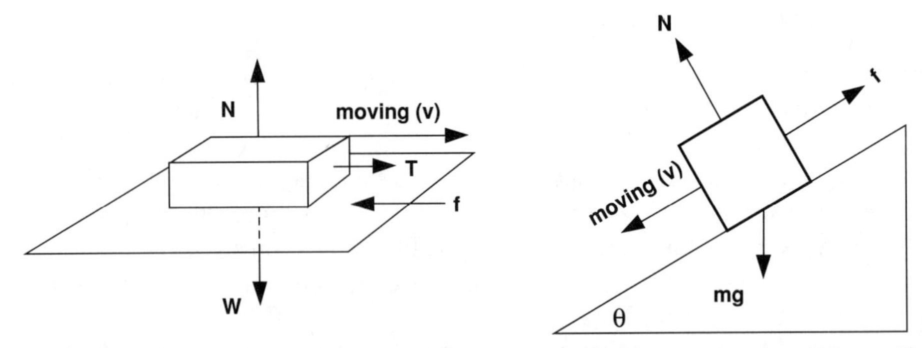

Figure 4-2. Forces acting on a block on a flat surface (left) and on an incline (right). For each situation showing a block with a velocity, note the direction of the normal force (N), weight (W or mg), and the friction force (f).

Example 4-1

A 4-KG BLOCK rests on a frictionless surface and is subjected to a 10-Newton force to the right. Find the acceleration of the block.

SOLUTION

Start by drawing a diagram (Figure 4-3).

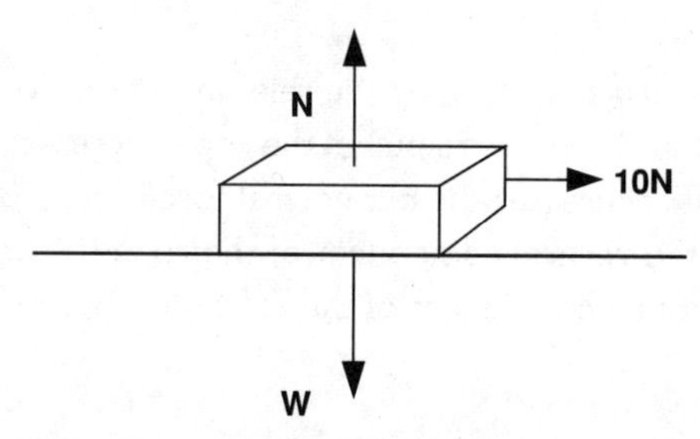

Figure 4-3. Diagram of relevant forces in Example 4-1.

The block has weight and a normal force, but no friction. Therefore, these forces do not affect the motion to the right.

Newton's second law:

$$F = ma$$
$$10\ N = (4\ kg)(a)$$
$$a = 2.5\ m/sec^2$$

Example 4-2

FOR THE PROBLEM detailed in example 4-1, find the acceleration of the block if the surface has a coefficient of kinetic friction of 0.2.

SOLUTION

Draw a diagram showing all the force vectors. Because this situation involves friction, remember to direct the friction force vector opposite to the applied force vector (Figure 4-4).

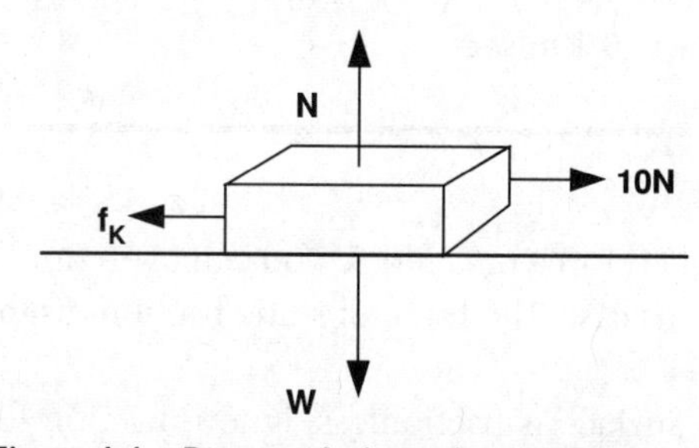

Figure 4-4. Diagram of relevant forces in Example 4-2.

Newton's second law:

$$F = ma$$
$$F_{(applied)} - f_k = ma$$
$$10\ N - \mu N = (4)(a)$$
$$N_{(normal\ force)} = W, \quad \text{because } \sum F_y = 0.$$
$$10\ N - 0.2(40\ N) = (4)(a)$$
$$a = 0.5\ m/sec^2$$

EXAMPLE 4-3

A 40-KG BLOCK is pulled to the right by a massless rope with a force of 100 N at an angle of 45° to the horizontal. If the coefficient of kinetic friction is 0.2, find the acceleration of the block.

SOLUTION

Figure 4-5 is a vector diagram showing the relevant forces. The most important part of the problem is in the setup. The upward pull of the rope decreases the friction of the block with the surface; it actually decreases the net normal force acting on the block. As shown in the following equation, the vertical component of the rope has been subtracted from the normal force. The horizontal component of the rope pull imparts the force that pulls the block to the right.

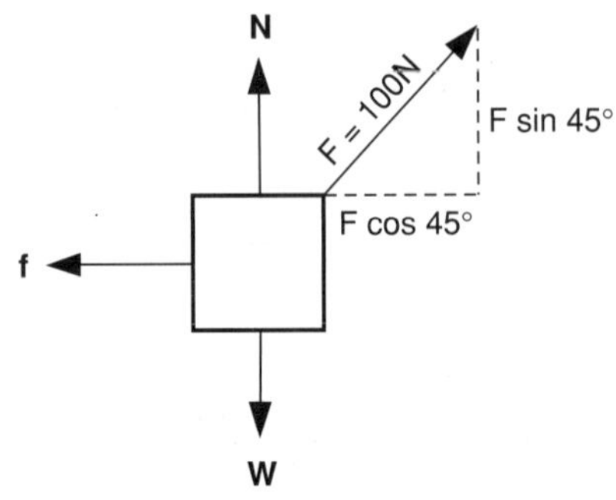

Figure 4-5. Diagram of the relevant forces and both *x* and *y* components of the upward rope pull in Example 4-3.

$$F = ma$$
$$F\text{ (x-component)} - \mu N = (40\text{ kg})(a)$$
$$N\text{ (normal force)} = w - F\sin 45°\text{, because } \sum F_y = 0.$$
$$100\cos 45° - (0.2)(mg - 100\sin 45°) = 40a$$
$$(100)(0.71) - (0.2)[400\text{ N} - 100(0.71)] = 40a$$
$$a = 0.1\text{ m/sec}^2$$

By increasing the applied force on a block, you can overcome the force of static friction and start the block moving. In this case, the force of static friction disappears and the force of kinetic friction takes over.

If you are told that a surface is frictionless, ignore friction forces completely and concentrate on the applied forces. Consider the following examples of these concepts.

IV. Incline Problems

Problems that involve inclines or ramps are just variations of problems you have already reviewed. Figure 4-6 shows the derivation of the various force vectors based on similar triangles. Note that the force vector that is parallel to the plane of the ramp is the **mgsinθ** term, and the force vector that is perpendicular to the ramp is the **mgcosθ** term.

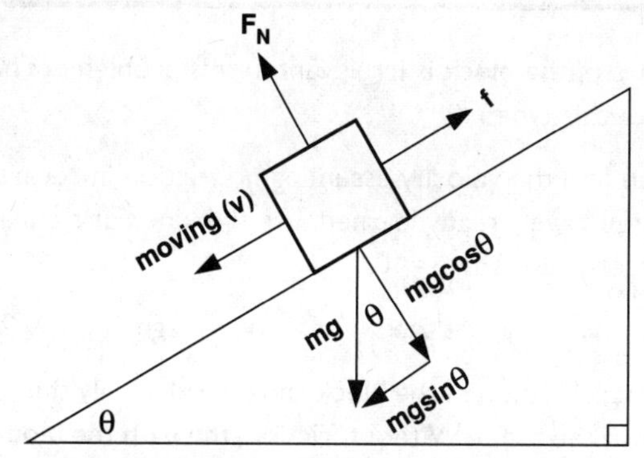

Figure 4-6. Schematic incline plane and its forces.

Example 4-4

A BLOCK OF MASS 10 kg slides down a 100-m long ramp that makes an angle of 30° to the horizontal. The ramp is frictionless. Find the acceleration of the block. Assume $g = 10\ \text{m/sec}^2$.

SOLUTION

The only force acting on the block in the direction of motion is the force vector directed down the ramp, $mg\sin\theta$.

$$F = ma$$
$$mg\sin 30° = ma$$
$$(10\ \text{m/sec}^2)(0.5) = a$$
$$a = 5\ \text{m/sec}^2$$

Example 4-5

A BLOCK OF UNKNOWN mass slides down a 100-m long ramp with a coefficient of kinetic friction of 0.1. The ramp makes a 30° angle to the horizontal. Assume $g = 10\ \text{m/sec}^2$.

1. Find the acceleration of the block.
2. Find the velocity at the bottom of the ramp if the ramp becomes frictionless.

SOLUTION

1. Acceleration (a) of the block is as follows:

$$F = ma$$
$$mg\sin\theta - \mu N = ma$$
$$mg\sin\theta - (0.1)(mg\cos\theta) = ma$$
$$m(10\ \text{m/sec}^2)(\sin 30°) - (0.1)(m)(10\ \text{m/sec}^2)(\cos 30°) = ma$$
$$a = 4.13\ \text{m/sec}^2$$

more

Ex. 4-5 cont.

Notice that the mass of the block is irrelevant in this problem. (The mass term cancels out in many problems of this type.)

2. You can find the velocity assuming no friction in several ways. To apply a technique you have already learned, use the kinematic equations. Assume the block starts from rest, so $v_o = 0$.

$$v_f^2 - v_o^2 = 2ax \quad \text{or} \quad v_f = (2gh)^{1/2} = \sqrt{2gh}$$

Use gravity for the acceleration of the block and consider only the vertical distance the block falls as it moves down the ramp. Without friction, the path the block takes to cover this vertical fall is independent of the exact path. This equation relates to conservation of energy.

Use the 30-60-90 rule to find the height of the triangular ramp (Figure 4-7):

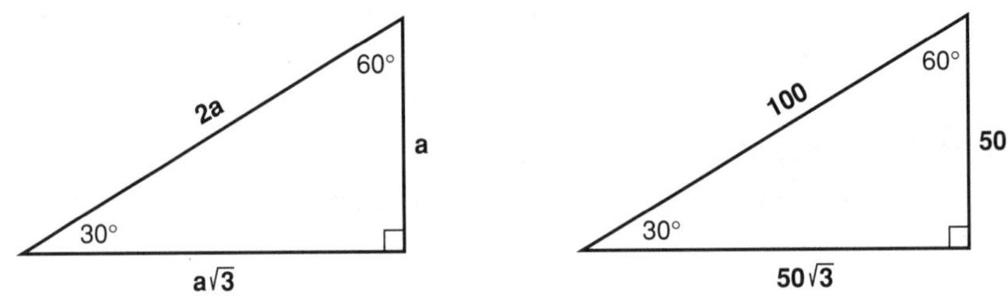

Figure 4-7. Use of the 30-60-90 triangle to solve for an unknown length in Example 4-5.

Thus, h = 50 m and the acceleration is equal to g. Thus, the solution is:

$$v_f = \{(2)(10\ \text{m/sec}^2)(50\ \text{m})\}^{1/2} = 32\ \text{m/sec}$$

Solve problems involving incline planes in the same way you did those involving flat surfaces: with the principles of Newton's second law. Include all forces that act on the movement of the object and set these equal to mass times acceleration.

V. Systems of Connected Bodies

Newton's laws also apply to a group of objects that are connected to one another.

Example 4-6

FIND THE ACCELERATION of the system in Figure 4-8. Assume that block m_1 has a mass of 2 kg, block m_2 is 4 kg, and block m_3 is 5 kg. The pulleys and the tabletop are frictionless.

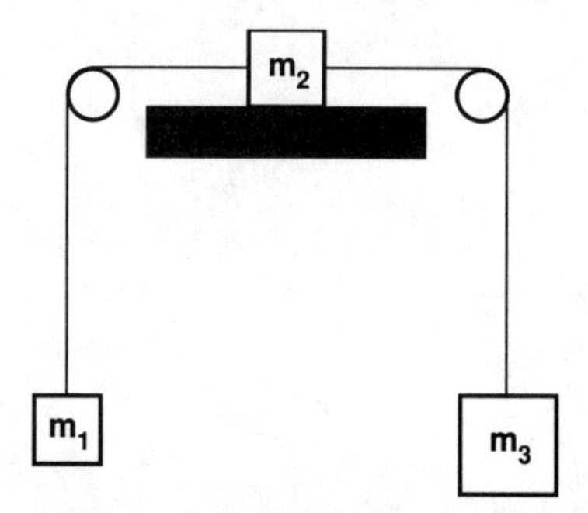

Figure 4-8. Diagram for Example 4-6.

SOLUTION

Draw a vector diagram that demonstrates the involved forces. Note the net movement to the right because $m_3 > m_1$ (Figure 4-9).

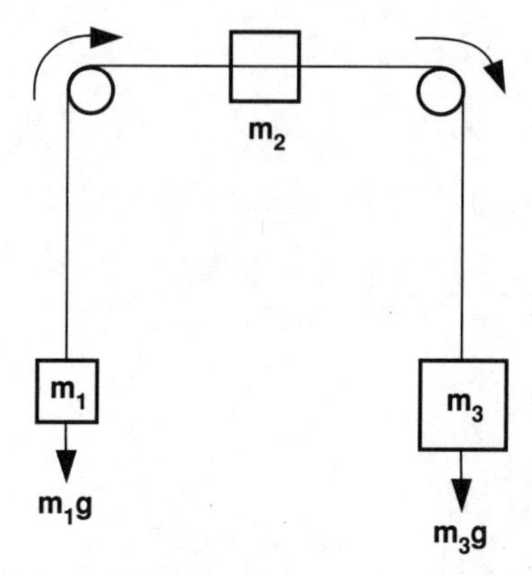

Figure 4-9. Vector diagram demonstrating the involved forces in Example 4-6. Note movement is to the right because $m_3 > m_1$.

You can see that the system will move to the right so that m_2 moves toward m_3. Set up an equation describing all the forces, and set it equal to the total system mass times system acceleration.

$$\begin{aligned}
\text{(forces to the right)} - \text{(forces to the left)} &= \text{(total system mass)(acceleration)} \\
m_3g - m_1g &= (m_1 + m_2 + m_3)(a) \\
(5\text{ kg})(10\text{ m/sec}^2) - (2\text{ kg})(10\text{ m/sec}^2) &= (11\text{ kg})(a) \\
50\text{ N} - (20\text{N}) &= 11a \\
a &= 2.7\text{ m/sec}^2
\end{aligned}$$

How could you set up this problem if there was friction? Because the system moves to the right, friction force would act to the left. Simply add friction force to the (forces to the left) portion of the preceding equation.

ergy work power

5 Work, Energy, and Power

This chapter is a review of the basic concepts, key definitions, and examples of work, energy, conservation of energy, efficiency, and power.

I. Work

Work is the product of the magnitude of a force and the distance over which the force acts. Work may also be thought of as the product of a force magnitude acting on an object and the object's displacement(s). There must be a net displacement of the object over which the force acts (Figure 5-1).

To calculate work, multiply the component of the force that is parallel to the object's displacement(s) to the magnitude of the displacement.

If the force direction and the direction of displacement are not parallel or are in the same direction, simply use components. If the angle θ is the angle between the force vector and the direction of displacement, then $\cos\theta$ is the component of this angle that is parallel to these two vectors.

Thus, the definition of work is:

$$\mathbf{W = Fs} \qquad \text{or} \qquad \mathbf{W = Fs\cos\theta}$$

The unit of work in the SI system is the Newton-meter (Nm), which is also known as the joule (J). A joule is the work done when an object is moved 1 meter against an opposing force of 1 Newton. In the British system, the unit of work is the foot-pound (ft-lb).

II. Conservation of Energy

A. ENERGY

Energy topics in physics usually relate to the storage of energy based on position or motion. Two basic types of energy are **potential and kinetic energy.**

Potential energy is the amount of work "stored" in an object based on its position, shape, or configuration relative to other objects. Several types of potential energy are defined as follows.

Gravitational potential energy relates to the "storage" of energy when an object is raised above an arbitrary reference point. The higher you raise an object, the greater its potential energy. To find the gravitational potential energy, use the equation:

$$\mathbf{PE = mgh}$$

in which h is the height from the reference point.

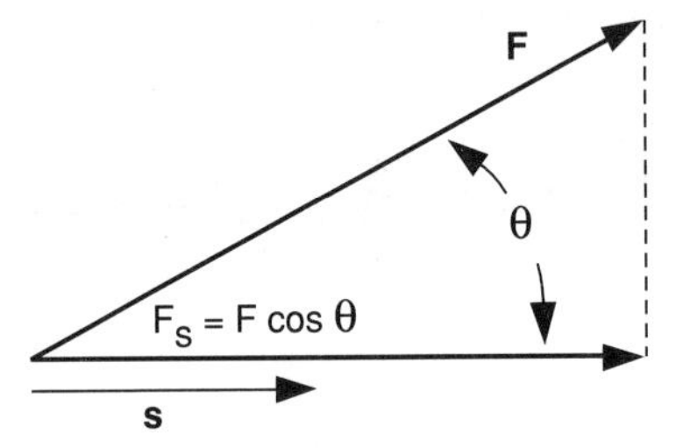

Figure 5-1. Concept of work. If F_s is the component of F along s, then $F_s = F\cos\theta$. Work is then $W = Fs\cos\theta$.

A second type of potential energy is the potential energy of a spring. When a spring is stretched or compressed from its resting position, energy is stored in the form of elastic recoil. To calculate this potential energy, use the equation:

$$\textbf{PE of spring} = \mathbf{1/2\ kx^2}$$

in which k is the spring constant and x is the distance of the spring from its equilibrium or resting point.

The spring constant is a property of the spring. The stiffer a spring, the greater its spring constant. As shown in Figure 5-2, x is the additional stretch or compression of a spring from its equilibrium point. A stiffer spring has a higher value of k and more energy associated with it—and it is harder to compress. The greater the value of x, the greater the energy of recoil.

Kinetic energy is that energy associated with the motion of an object. A moving object has the capacity to do work on another object, and this ability relates to the mass and velocity of the moving object.

$$\mathbf{KE = 1/2\ mv^2}$$

Example 5-1

A BOY IN A CART is pulled by a rope attached to the front of the cart. If the applied force is 10 N and the cart is pulled 10 m, answer the following questions:

1. Find the work done.

   Solution: $W = Fs = (10\ N)(10\ m) = 100\ J$

2. Find the work done if the rope makes an angle to the front of the cart of 60° and the cart continues traveling straight ahead as in question 1.

   Solution: $W = Fs\cos\theta = (10\ N)(10\ m)(\cos 60°) = 50\ J$

3. Find the work done if the angle in question 2 becomes 90°.

   Solution: $W = Fs\cos\theta = (10\ N)(10\ m)(\cos 90°) = 0$

Note: This example shows that if the line of action of a force is at a right angle to the displacement, no work is done.

Example 5-2

A MAN PERFORMS the following two actions with a book. Determine whether he does or does not perform work with respect to the book in each scenario given.

Scenario 1. A man holds a book at the level of his waist and then lifts the book over his head.

Scenario 2. A man holds a book at waist level while walking steadily.

SOLUTION

In each scenario, think about in what direction the applied force is directed. The man supports the book against the book's weight, and an upward force is applied on the book. In scenario 1, displacement of the book is in the same direction or parallel to this upward force. Thus, work is done.

In scenario 2, the displacement of the book is at a 90° angle to the upward force and no work is done. Note also that the man "is walking steadily," which suggests no acceleration. The fact that scenario 2 is not associated with work may be against your common sense because energy is expended in performing the actions described in this scenario. This question is only about work, however, and you must follow the conceptual definition of work to answer this and similar questions.

B. CONSERVATION OF ENERGY

Conservation of energy says that energy of a system cannot be created or destroyed, only converted from one kind to another. A ball raised above the level of the ground is gaining potential energy. When the ball is dropped, the potential energy is converted to kinetic energy. This kinetic energy is converted to heat as the ball is subjected to air friction and friction with the ground. All forms of energy are thus interconverted and not lost.

If friction is absent, the sum of the potential and kinetic energies of a system before an event will be equivalent to the sum of potential and kinetic energies after an event. Consider an example in which a cart is pushed up a hill. The energy during the push up the hill is primarily kinetic, yet the higher the cart goes up the hill, the more kinetic energy is converted to potential energy. Finally, when the cart is at the top of the hill, all the kinetic energy associated with the cart's movement has been converted to potential energy. The magnitude of the kinetic energy expended should ideally equal the potential energy stored if friction is absent.

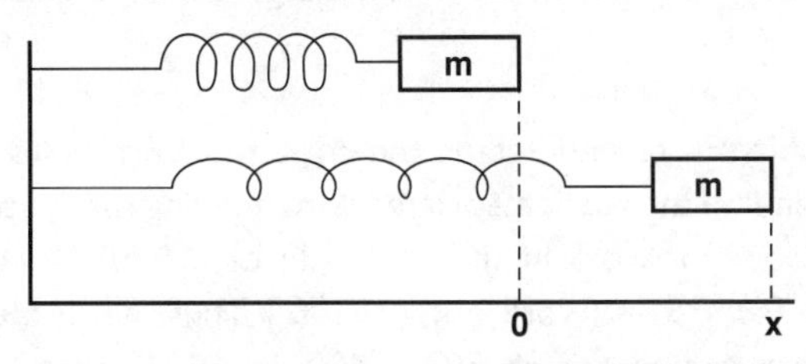

Figure 5-2. A mass-spring system. Note that x is the distance a spring is displaced from its equilibrium or resting position.

EXAMPLE 5-3

A ROLLER COASTER CART starts at rest from the top of a crest 5 m high. Find the velocity of the cart at the base of the crest, assuming a frictionless surface.

SOLUTION

Use conservation of energy.

$$\begin{aligned} PE + KE \text{ (top)} &= PE + KE \text{ (bottom)} \\ mgh + 0 &= 0 + 1/2mv^2 \text{ (Note: masses cancel)} \\ (10 \text{ m/sec}^2)(5 \text{ m}) &= (1/2)(v^2) \\ v &= 10 \text{ m/sec} \end{aligned}$$

A common sense expression for the conservation of energy follows:

$$\textbf{Conservation of Energy: } PE_a + KE_a = PE_b + KE_b$$

in which a and b represent two points (no friction).

III. Power

Power is the rate at which work is done. The unit of power in the SI system is the watt (W). The watt equals 1 joule per second.

$$\textbf{Power} = \textbf{Work/time}$$

assuming applied force is constant.

EXAMPLE 5-4

A ROPE/PULLEY SYSTEM is used to hoist a block. The pulley system has an element of friction and, as a machine, acts with 80% efficiency. If 10 N of force is applied to the system and the rope is pulled 50 m, determine how much energy is lost as friction/heat.

SOLUTION

The key to this problem is to understand the conceptual principles. If the pulley system is 80% efficient, you can find the work associated with hoisting the block. The work associated with pulling on the rope, or work input, is $W = Fs$ or (10 N)(50 m) = 500 J. Because the efficiency is 80% or 0.8, 0.8 = (work output)/500 J. Thus, work output must equal 400 J. Some energy is lost as friction, namely 500 J − 400 J = 100 J.

A machine is a mechanical device that transmits changes in the magnitude or direction of an applied force. Common examples include pulleys, levers, incline planes, and gears.

Efficiency is work output of a machine divided by work input. This fraction is usually represented as a percent.

Mechanical advantage is the ratio of output force to input force.

Conceptual Questions

1. A man walks up a steep, 500-m long hill. He decides to walk in a zigzag path, which is 500 meters long. How does the energy he expends and the force he exerts for his zigzag path compare to the straight hill climb?

   Solution: The energy expended in each case is the same, because all paths to the top require the same energy output. Energy, in this case, is work: W = Fs. The work to get to the top of the hill is the same for both paths, but the distance is not the same for both paths. If the distance is doubled, the force is cut in half. Thus, the zigzag path is associated with the same energy expenditure but only one half the force exerted.

2. A cart starting from rest rolls down a hill and reaches 10 m/sec at the bottom. If the same cart starts rolling down the hill from an initial speed of 5 m/sec, will its speed at the bottom be less than, equal to, or greater than 15 m/sec?

   Solution: The final speed would be less than 15 m/sec. Rolling down the hill adds a certain amount of kinetic energy, but not a certain amount of speed. Speeds do not add in this problem because each cart spends a different amount of time on the hill in which to gain speed. When the cart starts at 5 m/sec, it spends less time on the hill and so picks up less speed going down. To make the calculations in this problem, convert speed to kinetic energy and add energy units.

Momentum and Impulse 6

Momentum is a conserved quantity. When two objects collide, the momentum of each may change, but the total momentum of the system remains constant. A clear understanding of this property is extremely useful.

I. Momentum and Impulse

A. MOMENTUM (p)

Momentum is the product of the mass and velocity of an object. Momentum allows you to analyze motion in terms of the mass and velocity of an object rather than its force and acceleration. Momentum is a vector that has the same direction as the object's velocity.

$$\mathbf{p} = \mathbf{mv}$$

B. IMPULSE

Impulse is a force multiplied by the time during which the force acts. An impulse is the product of a force applied over a short time causing a change of momentum. A body receives momentum by the application of an impulse.

$$\text{Impulse} = Ft = \Delta\, mv$$

II. Conservation of Momentum

Conservation of momentum states that if the net force acting on a system is zero, the total linear momentum of the system will remain constant. Therefore, the momentum of the bodies before a collision equals the momentum of the bodies after the collision.

$$p_1 + p_2 = p_1 + p_2$$
$$\text{Before} = \text{After}$$

The momentum of a grenade before its explosion equals the sum of the momentum of all the fragments of the grenade after its explosion. The velocity vectors of the fragments cancel each other out when all the momentum vectors are added together.

III. Elastic and Inelastic Collisions

A. ELASTIC COLLISIONS

Elastic collisions occur between two or more bodies in which no kinetic energy is lost and the total linear momentum is constant.

An example of an elastic collision is when two balls on a pool table strike one another. The sum of the momentum of the balls before the collision equals the momentum after the collision. Also, the sum of the kinetic energies of the balls before contact equals the sum after contact.

Elastic Collision = Momentum conserved, kinetic energy conserved

B. INELASTIC COLLISIONS

Inelastic collisions occur between two or more bodies in which kinetic energy is lost because of transformation to heat, sound, and the like. Momentum of the bodies before and after the collision is constant. The collision is **completely inelastic** if the colliding particles stick together after the collision.

EXAMPLE 6-1

A 1-KG BLOCK sits at rest on a frictionless surface. A smaller block, moving to the right with mass 0.5 kg, strikes the 1-kg block with a speed of 2 m/sec. Assuming a perfectly elastic collision without friction, find the velocities of the two blocks after the collision.

SOLUTION

In the following equations, the smaller block is given the subscript 1, and the larger block is given the subscript 2. Before collision is designated by "i" for initial, and after the collision is denoted by "f" for final.

The conservation of momentum equation is:

$$\begin{aligned} p_{1i} + p_{2i} &= p_{1f} + p_{2f} \\ m_1 v_{1i} + 0 &= m_1 v_{1f} + m_2 v_{2f} \\ (0.5\ \text{kg})(2\ \text{m/sec}) &= (0.5\ \text{kg})(v_{1f}) + (1\ \text{kg})(v_{2f}) \\ 1 &= 0.5\, v_{1f} + v_{2f} \end{aligned}$$

This equation has two unknowns, which means the conservation of energy equation is as follows:

$$KE_i + PE_i = KE_f + PE_f$$

Given only KE in this problem:

$$\begin{aligned} 1/2\, m_1 (v_{1i})^2 &= 1/2\, m_1 (v_{1f})^2 + 1/2\, m_2 (v_{2f})^2 \\ 1/2(0.5\ \text{kg})(2\ \text{m/sec})^2 &= 1/2(0.5\ \text{kg})(v_{1f})^2 + 1/2(1\ \text{kg})(v_{2f})^2 \\ 1 &= 1/4(v_{1f})^2 + 1/2(v_{2f})^2 \end{aligned}$$

To solve for the final velocities, take the final equations from the conservation of momentum and conservation of energy expressions and solve for the unknowns. You can use any algebraic method. If you solve these equations, you find that $v_{2f} = 4/3$ m/sec and $v_{1f} = -2/3$ m/sec. The negative sign going with the final velocity of block 1 means that this block moves in the opposite direction from which it came. Thus, this block moves to the left after the collision.

A good example of an inelastic collision is two cars colliding at high speed. The energy associated with the kinetic energy of the cars is transformed to heat deformation and sound as the cars collide. The smashed wreck of the two colliding cars has a momentum equal to the sum of the precrash momentum of the cars, assuming no friction with the ground.

Inelastic Collision = Momentum conserved, kinetic energy not conserved

Example 6-2

A CANNON IS BOLTED to the floor of a closed railroad car (Figure 6-1). The railroad car sits on a frictionless track at rest and has a mass of 10,000 kg. A cannonball of mass 10 kg is fired from the cannon with a velocity of 100 m/sec to the right.

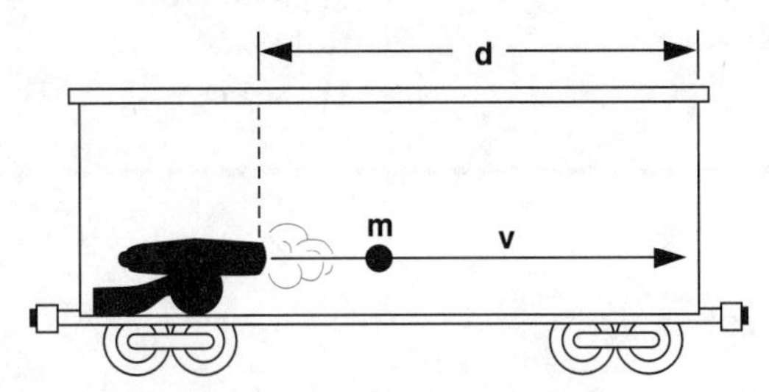

Figure 6-1. Diagram for Example 6-2.

1. Find the velocity and direction of any motion exerted on the railroad car.
2. When the cannonball strikes the far wall of the railroad car, it becomes embedded. What happens to the velocity of the railroad car when the cannonball strikes the wall?

SOLUTION

1. When the ball is fired to the right, the cannon is pushed to the left. Because the cannon is bolted to the floor of the railroad car, the entire cannon-railroad car complex is pushed to the left. Because the mass of this complex is great, its velocity of motion will be small. The momentum before the firing is zero, so after firing the momentum of the ball to the right must be equal in magnitude to that of the car and cannon to the left. Thus, $mv = MV$.

$$V = mv/M$$

or

$$V = (10\text{ kg})(100\text{ m/sec})/(10{,}000\text{ kg}) = -0.1\text{ m/sec}$$

Note: The negative sign shows that the velocity vector of the railroad car is opposite that of the cannonball.

2. As the cannonball becomes embedded in the wall, it exerts a force on the wall to the right. The wall exerts an equal and opposite reaction force on the ball to the left. The ball and railroad car both stop moving because the net momentum is still zero. Thus, the railroad car will roll to the left while the ball is in the air. Once the ball strikes the wall, the car will stop moving.

EXAMPLE 6-3

A 100-g BIRD flying at 20 m/sec collides with a hovering, stationary, 10-g hummingbird in the air such that the two birds stick together after the collision. Assuming a one-dimensional collision, find the common speed of the two entangled birds just after the collision.

SOLUTION

This problem illustrates an inelastic collision, albeit much simplified because only one dimension need be considered (no angles). If the birds were flying at angles to one another, you would have to figure out components of their velocities.

Conservation of momentum:

$$mv_i + mv_i = mv_f$$
$$(0.1\ kg)(20\ m/sec) + (0.01 kg)(0\ m/sec) = (0.11\ kg)(v_f)$$
$$2 = 0.11\ v_f$$
$$v_f = 18\ m/sec$$

Conceptual Questions:

1. A continuous force acts on an ice skater on a friction-free ice skating rink that causes her to accelerate. This applied force causes her speed to increase a certain amount. What would happen to the speed of the ice skater if: (a) the force and mass of the skater are unchanged but the time the force acts is tripled, and (b) the force is doubled while the mass and action time are unchanged?

   Solution: Consider this question conceptually. The force described increases the speed of the skater a certain amount each second it acts. Thus, if you triple the time the force acts, you must triple the increase in speed. The magnitude of the force applied also can increase the speed. Without applied force, no speed change occurs. A small force produces a small speed change. If you double the force, you double the speed change by doubling the acceleration.

   You can also think about this problem mathematically. $Ft = \Delta mv$, so $\Delta v = Ft/m$. Force and time vary directly, while mass varies inversely to Δv.

2. The physics of riot control are discussed in a police academy class. Suppose that rubber bullets bounce off people, whereas lead bullets penetrate.

   Officers are told that rubber bullets are more effective than lead bullets in riot control because they are more likely to knock rioters to the ground and cause less tissue damage. On the basis of your understanding of basic mechanics, are both of these assertions correct if a rubber and a lead bullet have the same size, speed, and mass?

   Solution: Yes, these assertions are correct. The equal momentum values of both bullets as they leave a gun change as soon as they make contact with a rioter. The impulse of the rubber bullet is greater because it bounces back, whereas the lead bullet penetrates. The rubber bullet contacts the rioter for significantly more time, and impulse equals the product of force and time. The impulse is greater for the rubber bullet because the rioter provides not only the impulse to stop the rubber bullet, but also the additional impulse to cause rebound of the bullet. Depending on the elasticity of the rebound, up to twice the impulse for the impact of the rubber bullet with the rioter occurs, and therefore up to twice the momentum is imparted to the rioter. Remember $Ft = m\Delta v$. Thus, the rubber bullet is more likely to knock over the rioter. The momentum of the lead bullet is completely transferred to the rioter, who supplies the necessary impulse to stop it.

Although the rubber bullet gives the rioter the most momentum, it does not give the most energy. When the rubber bullet bounces back, it keeps much of its kinetic energy. The lead bullet slows and stops, surrendering most or all of its kinetic energy as heat and deformation-tissue damage.

In conclusion, the rubber bullet puts much momentum but little energy into the collision with the rioter, whereas the lead bullet puts in little momentum but much energy into the collision.

7 Circular Motion and Gravitation

This first of two chapters concerning rotational systems is an overview of the basics of circular motion and its application to the study of gravitation. Chapter 8 provides a review of the physics of rotational dynamics, including inertia and torques. This material is confusing, so concentrate on the basic conceptual principles.

I. Circular Motion

Figure 7-1 illustrates a point that rotates around a circle from point A to point B. As the object rotates, it creates an angle θ measured at the center of the circle. **The magnitude of the angle θ through which the object passes is the angular displacement.** The quantities (r) and (s) represent the radius and arc length of the objects rotation, respectively.

Angular displacement can be represented in degrees, revolutions, or radians. Use the units of radians in situations in which both linear and circular quantities appear. By definition, **a radian is the angle the arc length of which equals the radius of the circle.** The important relationships between radius, arc length, and θ are as follows:

$$\theta = s/r \qquad \text{or} \qquad s = r\theta \qquad (\theta \text{ in radians})$$

Suppose an object rotates along a unit circle (radius = 1) and does one complete rotation. A comparison of the units of rotation follows:

$$360 \text{ degrees} = 2\pi \text{ radians} = 1 \text{ revolution}$$

It is important to review some basic definitions and to understand the difference between the tangential and angular measurements.

Tangential velocity is the velocity of a mass moving through a particular point on a circular path. The direction of the velocity vector is **perpendicular** to the radius of the circle. Figure 7-2 shows the direction of a tangential velocity or tangential acceleration vector. Think about what happens if you swing a ball from the end of a string over your head and then suddenly let go. The ball and string heads off in a direction tangential or perpendicular to the original circular path.

Tangential acceleration is the acceleration associated with a mass moving through a particular point on a circular path. The direction of the acceleration vector is tangent to the circle and perpendicular to the radius.

Figure 7-3 shows that the direction of the angular velocity and acceleration of a rotating point are totally different than the tangential velocity and acceleration. **The angular velocity and angular acceleration vectors are directed upward/downward along the axis of rotation.** These directions were assigned based on the mathematics of circular motion, and are not

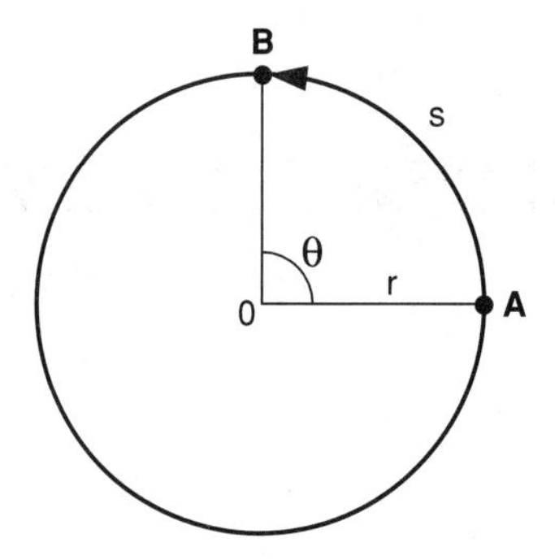

Figure 7-1. A unit circle. Note the radius (r), arc length (s), angular displacement (θ), and origin (O).

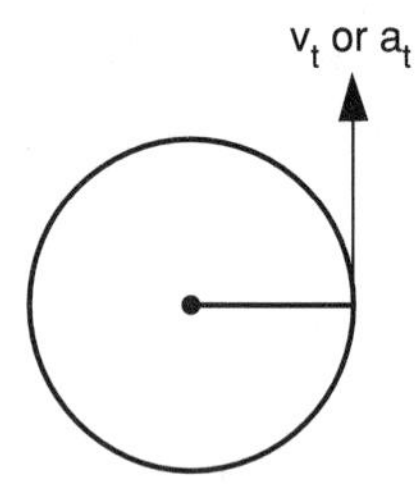

Figure 7-2. The tangential velocity or tangential acceleration for an object in a circular path is perpendicular to the radius of the circle.

directly intuitive.

Where does tangential acceleration arise? Consider a mass rotating in a circular path. If the angular velocity increases, so then does the tangential speed. Changes in angular speed translate into changes in tangential acceleration.

II. Angular Velocity and Acceleration

The average angular velocity of an object (ω or omega) is the change in angular displacement of an object divided by time required for the displacement to occur. The units are in radians per second.

$$\omega = \Delta\theta / \Delta t$$

Angular velocity in units of radians per second can also be found as follows: $\omega = 2\pi f$, in which f is frequency in revolutions per second.

The average **angular acceleration** of an object (α or alpha) is the change in angular ve-

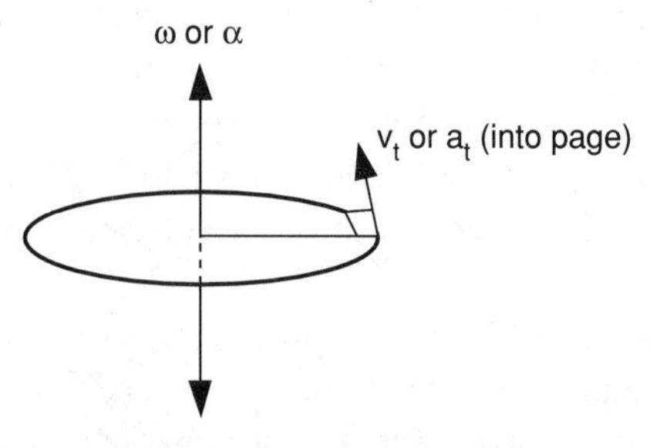

Figure 7-3. The direction of angular velocity and acceleration vectors are either upward or downward along the axis of rotation. Note that the tangential velocity or acceleration are directed tangent to the circle.

EXAMPLE 7-1

A BELT PASSES OVER a wheel with a radius of 25 cm. A point on the belt has a speed of 5 m/sec. Find the angular speed of the wheel.

SOLUTION

You are given the linear speed of the point on the belt (the units describe linear motion). Thus, $\omega = v/r$. $\omega = (5 \text{ m/sec})/(0.25 \text{ m}) = 20$ rad/sec. To convert radians per second into revolutions per second, multiply by the conversion factor:

$$(20 \text{ rad/sec})(1 \text{ rev})/(2\pi \text{ rad}) = 3.2 \text{ rev/sec}$$

locity of the object over the time interval during which this change takes place:

$$\alpha = \Delta\omega / \Delta t \qquad \text{or} \qquad \alpha = (\omega_f - \omega_o) / \Delta t$$

EXAMPLE 7-2

A ROPE WINDS AROUND a pulley with a radius of 5 cm. The pulley rotates at 30 revolutions per second and then slows uniformly to 20 revolutions per second over a time interval of 2 seconds. Answer the following questions.

1. Find the angular deceleration.
2. Find the number of revolutions associated with the two seconds of deceleration.
3. Find the length of rope that winds around the pulley in 2 seconds.

SOLUTION

1. $\alpha = (\omega_f - \omega_o) / \Delta t = (20 \text{ rev/sec} - 30 \text{ rev/sec})/2 \text{ sec} = -5 \text{ rev/sec}^2$
2. The number of revolutions should equal the average angular velocity multiplied by time.

   $$\theta = (\omega_{ave})(t) = 1/2(\omega_f + \omega_o)(t) = 1/2(20 + 30)(2) = 50 \text{ revolutions}$$

3. If you find the arc length around which the rope winds, you have an approximation of the rope length. Knowing that the pulley makes 50 revolutions in this time period enables you find the value of s.

   Convert revolutions into radians:

   $$(50 \text{ revolutions})(2\pi \text{ radians/1 revolution}) = 314 \text{ radians}$$

   Now, $s = \theta r = (314 \text{ rad})(0.05 \text{ m}) = 15.7$ m

   $$(\# \text{ revs})(2\pi r) = (50 \text{ revs})(2)(3.14)(0.05 \text{ m}) = 15.7 \text{ m}$$

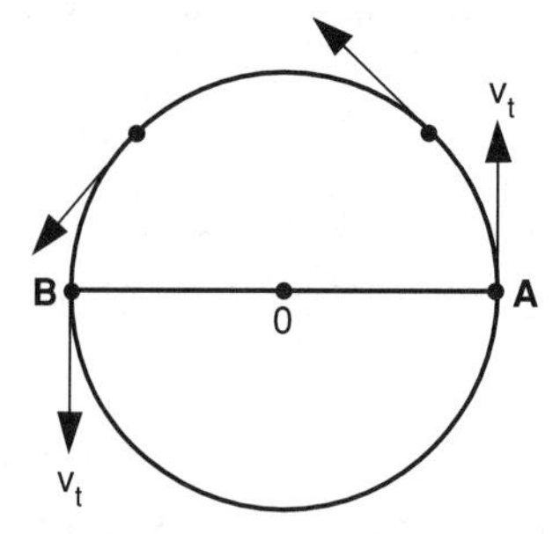

Figure 7-4. A point rotates in a circular path from A to B. Even at constant speed, the direction of the tangential velocity changes, which gives rise to a centripetal acceleration.

Linear velocity and accelerations can be converted into angular velocity and acceleration by dividing the linear quantities by the radius of rotation of the object:

$$\omega = \mathbf{v/r} \qquad \textbf{and} \qquad \alpha = \mathbf{a_t/r}$$

in which a_t is the tangential acceleration.

The vector directions of both angular velocity and acceleration were shown in Figure 7-3.

III. Centripetal Acceleration and Force

An object travels in a circular path at a constant rate. As the object rotates from A to B, its direction is constantly changing (i.e., the direction of its tangential travel). Objects naturally want to leave a circular rotation to travel tangentially. If you tie a weight to the end of a rope and begin rotating the rope and weight above your head in a circular path, you feel a force on your hand wanting to move the weight away from the circle. The force of your hand holding the rope resists the rope and weight leaving their circular orbit. If you let go of the rope, the weight would leave the circular orbit tangential to the spot on the circle from which it was released.

Study Figure 7-4. As a point rotates from A to B with constant speed, the direction of the tangential velocity vector changes. An acceleration is associated with this change because the velocity is changing over time. Velocity may change if direction changes and magnitude is constant. The acceleration associated with this velocity change is known as **centripetal acceleration** and is directed radially inward. **This acceleration is associated with the force holding the object in a circular path.**

Two types of acceleration are shown in Figure 7-5. Tangential acceleration (a_t) is the acceleration of a rotating point tangent to the circle. Centripetal acceleration (a_c) is the acceleration

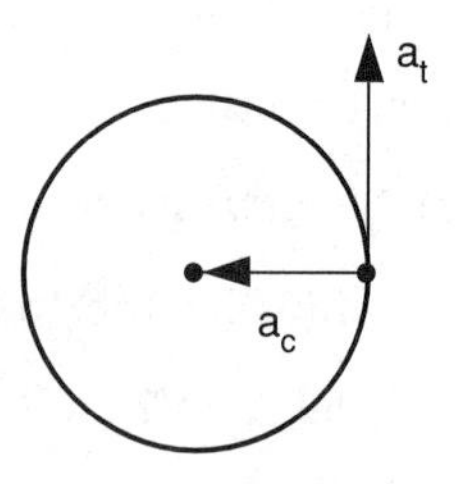

Figure 7-5. Centripetal acceleration and tangential acceleration vectors.

Example 7-3

ANSWER THESE QUESTIONS.

1. A car drives on a circular flat track at constant speed. What are the forces that act on the car and the road?

   Solution: The weight of the car acts on the road, and an equal and opposite force is imparted to the car from the road. A small backward force on the car results from air resistance, and a forward force is exerted by the road on the tires. Finally, an inward-directed friction force is exerted on the tires by the road.

2. Which force keeps the car moving in a circular fashion and prevents it from flying off the track?

   Solution: The friction force between the tires and the road prevent the car from leaving the track. A centripetal acceleration results from a frictional force exerted on the tires by the road. This frictional force acts perpendicular to the motion of the car and is directed inward, as is the centripetal acceleration.

3. What would make the car fly off the track?

   Solution: If the driver tries to negotiate a curve too rapidly, the maximum frictional force will be exceeded and the car will skid. If the tires are bald, a smaller maximal frictional force is all that can be achieved, and the car is at increased risk to skid off the track. A wet track surface will also interfere with the friction between the tires and the road.

associated with keeping an object rotating in a circular path and is directed inward. Use the following equation to find the magnitude of each type:

$$\mathbf{a_t = \alpha r}$$

Centripetal acceleration: $\mathbf{a_c = v^2/r}$ **or** $\mathbf{a_c = \omega^2 r}$

The force holding the rotating object in its circular path is the **centripetal force (F_c)**:

$$\mathbf{F_c = ma_c = mv^2/r = m\omega^2 r}$$

IV. Conceptual Relationships in Circular Motion

Examine the relationships between v, a_c, and a_t in Figures 7-6, 7-7, and 7-8. In each circle, assume that a point is rotating with the conditions given. Note that a_c is associated with a change in direction, whereas a_t is associated with a change in magnitude.

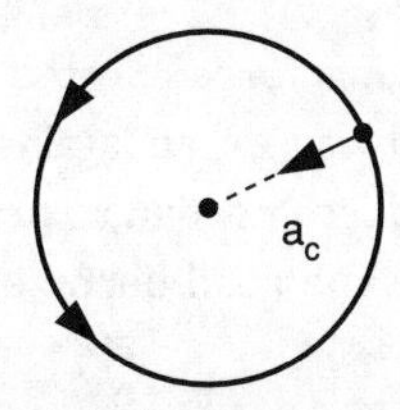

Figure 7-6. Uniform circular motion with constant speed. Note an acceleration, namely a_c, but no a_t, at constant speed.

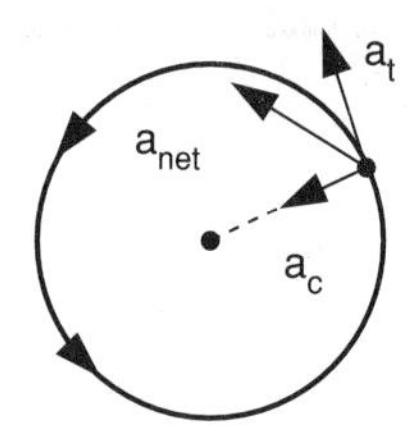

Figure 7-7. Ever increasing speed. Note an acceleration tangentially as well as centripetally in this situation. The net acceleration vector, therefore, is a combination of these forces.

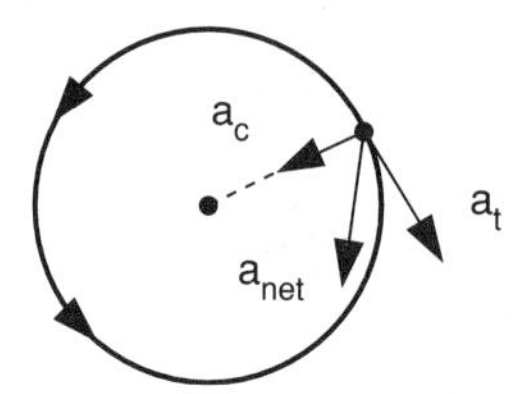

Figure 7-8. Ever decreasing speed. A deceleration vector is directed opposite the direction of motion. Again, the net acceleration is the result of adding a_c and a_t.

V. Gravitation

Centuries ago, Newton described the law of universal gravitation, which states that every object in the universe, regardless of its mass, attracts every other object with a force that is proportional to the masses of the two objects and inversely proportional to the square of the distance between them:

$$\mathbf{F = G(m_1\, m_2)/r^2}$$

$$G = \text{universal constant } (6.67 \times 10^{-11}\ \text{Nm}^2/\text{kg}^2)$$

The distance (r) is measured from the center of mass of one object to the center of mass of the other object.

Faraday developed the concept of a gravitational field, such that an object modifies the space surrounding it by establishing a gravitational field that extends outward in all directions. If a massless test point is placed near a second mass, such as the earth itself, you can derive the formula for the gravitational field of the earth (g):

$$\mathbf{g = GM/r^2}$$

Example 7-4

WHY DO YOU THINK that good highways have banked curves rather than flat curves?

SOLUTION

Banked curves ensure that the normal force exerted by the road on the car has a horizontal component—the force the highway exerts against the vehicle. This horizontal component can provide part of or all the force needed to produce the centripetal acceleration, reducing the role of frictional force. The road becomes safer, especially when road surfaces are slippery.

The gravitational field of a planet is related primarily to its mass. To find the value of g at the surface of a planet, simply use r = radius of the planet. For example, the value of g for the planet Jupiter would be:

$$g_{Jupiter} = G\,(M_{Jupiter} / r^2_{Jupiter})$$

Conceptual Questions:

1. Two identical objects rotate in circular paths of equal diameter. One object rotates twice as fast as the other. How does the centripetal force of the faster object compare to that of the slower object?

   Solution: The centripetal force varies as the square of the velocity or angular velocity. Thus, the centripetal force needed to keep the faster object rotating in a circular path is four times that needed for the slower object.

2. Two boys stand on opposite ends of a carousel that is turning clockwise. If one boy throws a ball directly toward the other, will the boy likely catch the ball?

   Solution: Most likely, the ball will not be caught. When initially thrown, the ball might start going toward the recipient. Once the ball is in the air, however, the recipient and carousel turn clockwise and the ball will miss its target. The ball carries with it the velocity imparted by the ball thrower. The principle that explains the deflection of objects subjected to turning is the **Coriolis effect.**

tational

8 Rotational Dynamics

dynamic

Mass is the property of an object that resists acceleration. The moment of inertia takes the place of mass in rotational motion. **The moment of inertia is the quantitative angular measure of the property of an object that resists rotational acceleration.**

I. Moment of Inertia

The moment of inertia of a rigid body depends on the shape and axis of rotation of the body. Different formulas allow calculation of the moment of inertia of various different shapes (Table 8-1). The formula for a point mass is important to know.

II. Rotational Torque and Angular Momentum

Applying Newton's second law, F = ma, to rotational systems results in an expression for rotational torque (τ):

$$\tau = \mathbf{I}\alpha$$

The rotational torque of an object is the moment of inertia times the angular acceleration. When compared to Newton's second law, the moment of inertia term replaces mass and the angular acceleration replaces linear acceleration.

The angular momentum (L) of a rotating object is similar to the expression for linear motion. Instead of mass, use moment of inertia (I) and instead of linear velocity, use angular velocity (ω).

$$\mathbf{L} = \mathbf{I}\omega$$

Just like with linear momentum, angular momentum is conserved if the external torque acting on the system is zero. Conservation of angular momentum says:

$$\mathbf{I}\omega_{\text{(initial)}} = \mathbf{I}\omega_{\text{(final)}}$$

Kinetic energy is associated with rotational motion. This type of energy is rotational kinetic energy. It is derived from the standard kinetic energy expression, although mass is replaced by moment of inertia and linear velocity is replaced by angular velocity.

$$\textbf{Rotational KE} = \tfrac{1}{2}\mathrm{I}\omega^2$$

TABLE 8-1. Moment of Inertia Formulas for Common Objects.

Shape	Formula
Point mass or thin ring	$I = mr^2$
Solid cylinder	$I = \frac{1}{2} mr^2$
Solid sphere	$I = \frac{2}{5} mr^2$

m = mass of the object; r = radius of the object.

EXAMPLE 8-1

WHAT HAPPENS WHEN you sit on a spinning bar stool with your legs folded, and then outspread them?

SOLUTION

The rate of spin decreases. In other words, your angular velocity decreases. This slowing occurs because the moment of inertia increases when your legs are outspread (radius increases) and because angular momentum is conserved, angular velocity must decrease.

EXAMPLE 8-2

WHY DOES AN ice skater in a spin turn faster when she pulls in her arms?

SOLUTION

This increase occurs because she is decreasing her moment of inertia (her radius decreases). Because angular momentum is conserved, her angular velocity increases.

III. Circular Applications of Kinematic Equations

The kinematic equations reviewed in Chapter 3 can be fully applied to rotational systems by substituting the angular velocities and accelerations for the linear values. Distances are replaced with angle values. The rotational kinematic equations are presented in Table 8-2. If you know the linear forms of these equations, you need not memorize the rotational forms.

TABLE 8-2. Rotational Kinematic Equations

Linear	Angular
$v = v_0 + at$	$\omega = \omega_0 + \alpha t$
$v_{(ave)} = \frac{1}{2}(v_i + v_f)$	$\omega_{ave} = \frac{1}{2}(\omega_i + \omega_f)$
$v_f^2 = v_0^2 + 2as$	$\omega^2 = \omega_0^2 + 2\alpha\theta$
$s = v_0t + \frac{1}{2}at^2$	$\theta = \omega_0 t + \frac{1}{2}\alpha t^2$

EXAMPLE 8-3

A ROTATING DISK, starting from rest, reaches a speed of 5 revolutions in 10 seconds. The moment of inertia of the disk is $I = \frac{1}{2}mr^2$, in which m = 1 kg and r = 2 m.

1. Find the angular acceleration of the disk.
2. Find the angular speed of the disk at t = 11 seconds.
3. Find the rotational torque of the disk.

SOLUTION

1. The angular acceleration = $\alpha = \Delta\omega/\Delta t$ = (5 rev/sec − 0)/(10 sec) = *0.5 rev/sec²*
2. The angular speed $\omega = \omega_o + \alpha t$; thus ω = (5 rev/sec)(2π rad/rev) + (0.5 rev/sec)(2π rad/rev)(1sec) = 10π rad/sec + 1π rad/sec = 11π rad/sec ≈ *34 rad/sec*

Note that the initial angular speed was taken at t = 10 sec, so that at time 11 sec, only 1 sec had elapsed from the initial setting.

3. Rotational torque = $\tau = I\alpha$ = (1/2 mr²)(0.5 rev/sec)(2π rad/rev) = *6.2 Nm*

hanical

Mechanical Properties

9

propert

The properties of solids are important topics in the health sciences, especially in such medical specialties as orthopedics and sports medicine. The key points in the following discussion are density, specific gravity, stress, strain, and elasticity.

I. Important Terms

The **density** of an object (**ρ**) equals its mass divided by its volume.

$$\rho = m/v \qquad \text{(units in SI system} = \text{kg/m}^3\text{)}$$

The specific gravity of an object is its density divided by a standard density. For gases, the density of air is used as the standard density. For liquids and solids, the density of water is used as the standard. Thus, the specific gravity is just a comparison of the density of an object to the density of a standard substance.

SG (liquid) = density of a particular liquid/density of H_2O

The **elasticity** of an object is the property by which a body returns to its original size after deforming forces are removed.

II. Stress, Strain, and Moduli

A. STRESS

Stress is a measure of the strength of an agent causing a deformation. A stress is a force applied over a unit area.

$$\text{Stress} = F/A \qquad \text{(units in SI system} = \text{Pascals (Pa) or N/m}^2\text{)}$$

Objects subjected to forces change their shape and may fracture. The fractional change in size or shape is the **strain,** and the force per unit area producing the deformation is the **stress.** For small applied forces or torques, the stress and strain in a material are usually linearly related. **The proportionality constant relating stress and strain is Young's modulus for tensile strain.**

B. STRAIN

Strain is the fractional deformation on an object resulting from a stress applied on the object.

$$\textbf{Strain} = \Delta \textbf{ dimension/original dimension (unitless)}$$

Three types of strain show dimensional changes as a result of an applied stress:

1. **Tensile strain:** The change in length of an object divided by the object's original length.

$$\Delta L/L_o$$

2. **Compressional strain:** The change in the volume of an object divided by the original volume of the object.

$$\Delta V/V_o$$

3. **Shear strain**: The distance a surface is sheared divided by the width of the object being sheared (Figure 9-1).

$$\Delta x/L$$

C. MODULI

The **elastic modulus** is the quotient of stress and strain. It is a measure of how hard it is to compress or stretch an object.

$$\textbf{Elastic modulus} = \textbf{Stress/strain}$$

Three types of moduli measure the fraction of stress over strain. The first type is **Young's modulus.** It is the ratio of stress and strain when an object stretches. The second type is **bulk modulus.** It is the ratio of stress to strain when the volume of an object changes. The third type is **shear modulus,** which is the ratio of stress to strain when an object is subject to shearing.

Young's modulus: Stretching stress/stretching strain = $F/A/_{\Delta L/Lo}$

Bulk modulus: Volume stress/volume strain = $\Delta P/_{-\Delta V/V}$

Shear modulus: Shearing stress/shearing strain = $F/A/_{\Delta x/L}$

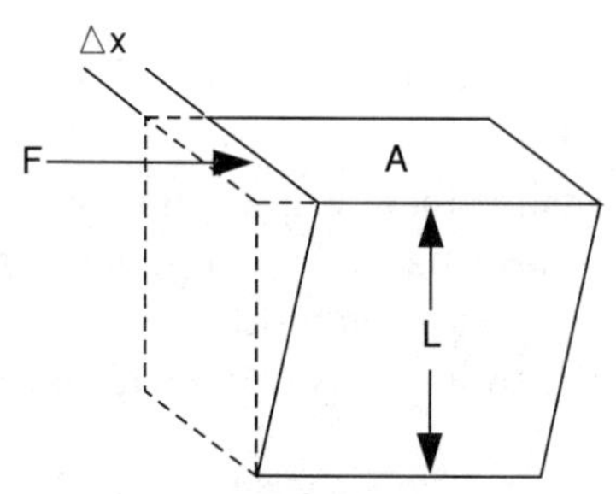

Figure 9-1. Shear strain. Note that a force causes the object to "shear" a distance (Δx). An example is pushing on the binding of a book. L represents the width of the object being sheared.

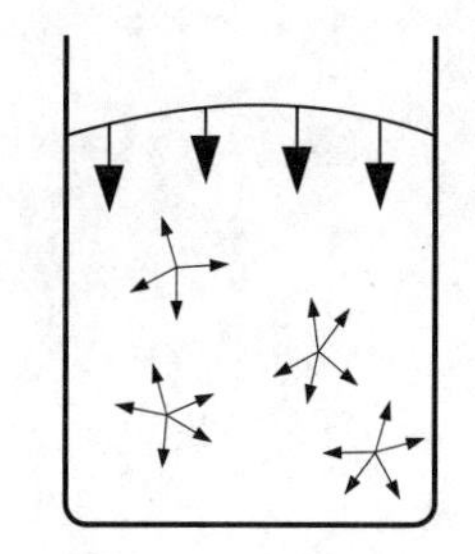

Figure 9-2. Surface tension.

III. Surface Tension

Surface tension is the amount of work (energy, or ΔE) required to expand the surface of a liquid by a given amount of area (ΔA).

$$\textbf{Surface tension} = \Delta E / \Delta A$$

Surface tension results from intermolecular attraction forces causing molecules on the surface of a liquid to be attracted to one another and to molecules below the surface. The molecules on the surface have a net attraction to the interior of the liquid, and try to form a surface that minimizes surface area. In Figure 9-2, notice that the surface of the liquid forms a "film" that minimizes surface area. An example is dewdrops you see early in the morning. These drops form spheres, which minimize the surface area for a given volume.

Cohesive forces are forces that attract like particles. An example is water molecules attracting one another. **Adhesive forces** are the attractions between unlike particles, such as water molecules being attracted to the glass wall of a container. If cohesive forces are larger than the adhesive forces in a given liquid, such as mercury, the fluid in a glass container will contact the container with a downward slope (Figure 9-3). If adhesive forces are stronger than cohesive forces, the liquid becomes attracted to the container sides and will slope upward to the point of contact (see Figure 9-3).

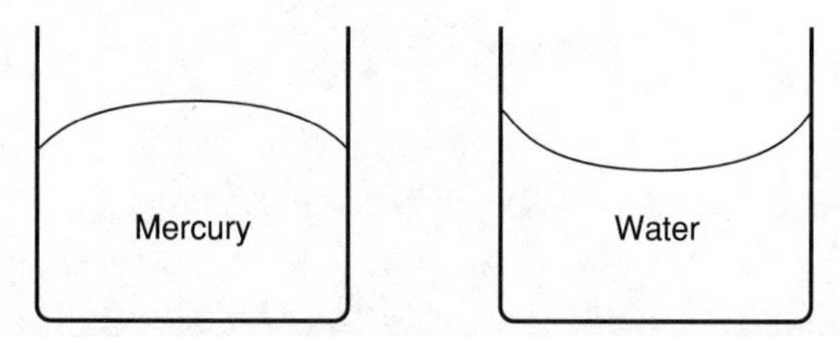

Figure 9-3. When cohesive forces are greater than adhesive forces, a downward sloping fluid may result, as in mercury (left). When adhesive forces are greater, an upward sloping fluid may result, as in water (right).

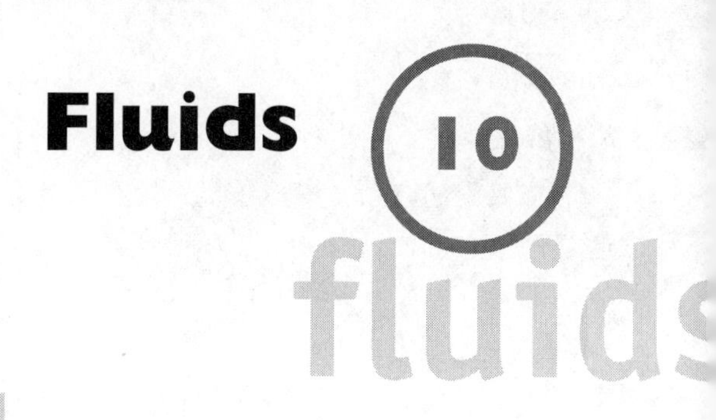

Fluids 10

Fluids are substances that have no rigidity, lack a fixed shape, and take the shape of the container that holds them. The focus of this chapter is on applying the laws of mechanics to fluids. Because a given mass of fluid does not have a fixed shape, the terms density and pressure are used instead of mass and force.

I. Fluid Statics

A. BASIC CONCEPTS

The **density** of a fluid (ρ) is the ratio of the mass of the fluid to its volume.

$$\rho = m/v \qquad \text{(in kg/m}^3\text{)}$$

The **pressure** exerted on a fluid is the ratio of the force acting on the fluid to the surface area at which the force acts:

$$P = F/A$$

Pressure is a scalar quantity and is measured in the SI system in units of N/m^2, or Pascals (Pa).

The pressure in a container of fluid increases as one moves deeper below the surface. You have experienced this principle in a swimming pool; the pressure on your eardrums is greater at a deep region in the pool than in a shallow region.

The **pressure at any depth** below the surface of a fluid is proportional to the fluid density and depth. It also depends on the acceleration of gravity (g), because the pressure at any depth depends on the weight of fluid over any point below the fluid surface.

To calculate the pressure at a point below the surface of a fluid (Figure 10-1), you must first remember that atmospheric pressure acts at the surface of a fluid. **Thus, the pressure at a point below the surface of a fluid will be the sum of the atmospheric pressure at the surface and the weight of the water above the point.**

Pressure at a depth (h) below the surface of a fluid:

$$P = P_{atm} + \rho gh$$

To find the pressure difference between two points in a fluid (Figure 10-2):

$$P_{diff} = \rho g \Delta h$$

in which Δh is the height difference between the two points.

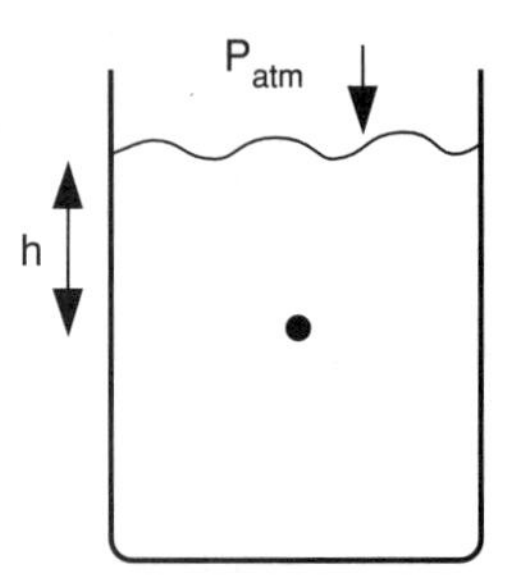

Figure 10-1. Pressure at a depth (h) below the surface of a fluid depends on the atmospheric pressure acting on the surface, the density of the fluid, the depth of the point of interest, and the acceleration of gravity.

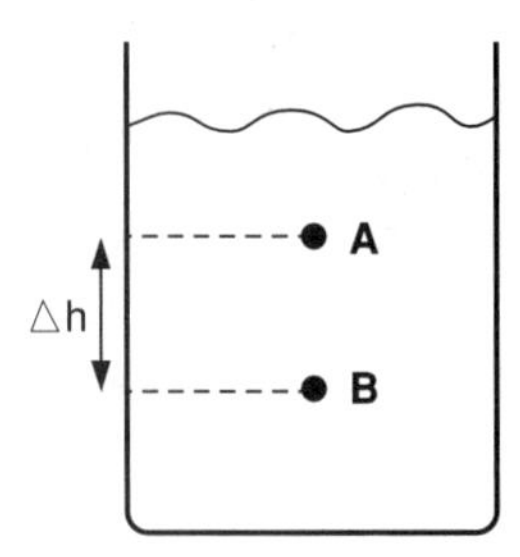

Figure 10-2. Difference in pressure between two points in a static fluid depends on the density of the fluid, the height difference between the two points, and the acceleration of gravity.

B. PASCAL'S PRINCIPLE

This basic law of fluids says that a change in pressure in a confined fluid is transmitted without change to all points in the fluid. Stated more simply, the pressure is the same at two places at the same depth in a fluid at rest. By Pascal's principle, points B, C, and D in Figure 10-3 are all subjected to the same pressure. Even if the container over point B was sealed at the fluid surface, the pressure at B would still equal that at C and D. Do not be tricked by the shape of the fluid container; the issue is the depth of the points in question.

The most common application of Pascal's principle is the **hydraulic lift** (Figure 10-4). A hydraulic lift contains an incompressible fluid. When a force is applied to the piston on one end of the lift, the pressure is transmitted through the liquid undiminished to the other piston on the opposite end of the lift. Therefore, a piston with small surface area transmitting a small force is effective in transmitting a large force to a large piston, because $P = F/A$ and the pressure is transmitted through the system unchanged. A similar situation occurs when you jack up your car. A force applied on a small surface-area piston generates pressure that is transmitted to the other end of the hydraulic jack. The pressure then acts on a larger surface-area piston, which, in turn, generates a larger force.

Thus, for hydraulic lifts or closed systems in which pressure is transmitted:

$$\mathbf{F_1/A_1 = F_2/A_2} \text{ or } \mathbf{F_2/F_1 = A_2/A_1} \text{ where } A_1 \text{ and } A_2 \text{ are the areas of the pistons}$$

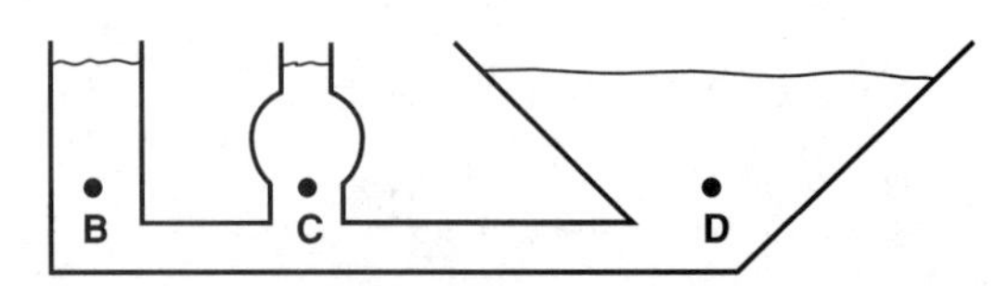

Figure 10-3. Pascal's principle. Pressure at points B, C, and D are equal.

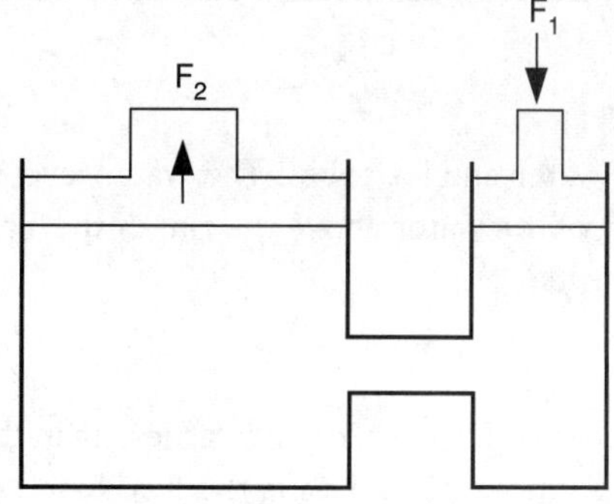

Figure 10-4. Hydraulic lift.

C. ARCHIMEDES' PRINCIPLE

An object that floats or is submerged in a fluid experiences an upward or buoyant force because of the fluid. The buoyant force is equal to the weight of fluid that the object displaces. **Archimedes' principle states that the buoyant force exerted on a body partly or completely immersed in a fluid is equal to the weight of the fluid displaced by the body.**

Figure 10-5 demonstrates that when an object with volume (V) is immersed in a fluid-filled container, several forces act on the object. The buoyant force (B) pulls the object to the surface of the fluid. The weight of the object (w_o) makes it sink in the fluid. Whether or not the object sinks or rises depends on the balance of these upward and downward forces.

To find the **magnitude of the buoyant force,** use the following formula:

$$\textbf{Buoyant force (B)} = \mathbf{mg} = \rho \mathbf{Vg}$$

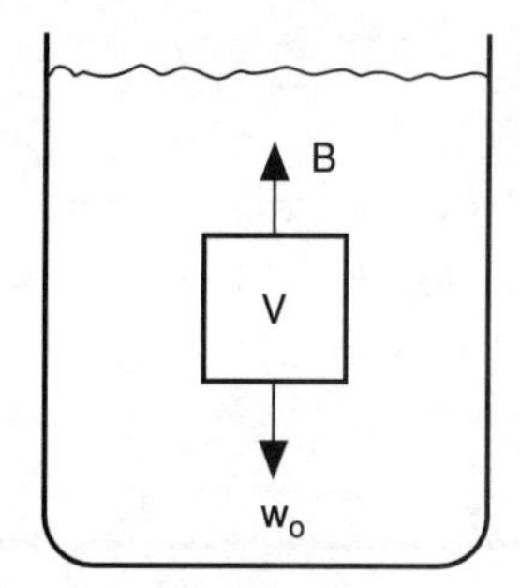

Figure 10-5. Archimedes' principle.

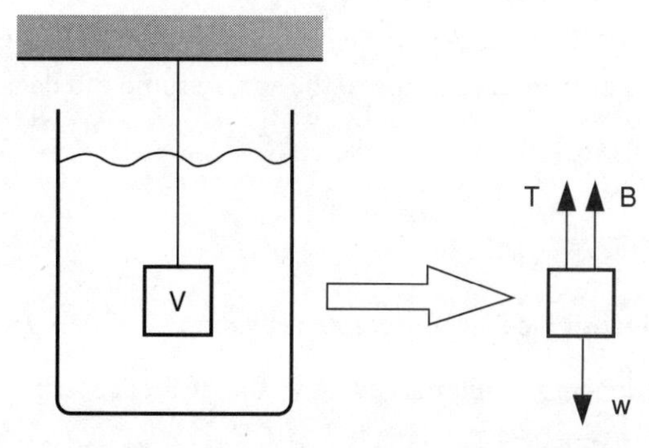

Figure 10-6. To find the tension on a string for the diagram in Figure 10-6, take the difference of the weight of the block and the buoyant force acting on the block.

EXAMPLE 10-1

A GLASS IS FILLED with water and ice cubes. The water level reaches the rim of the glass and the ice cubes rise out of the water above the rim of the glass. What would you expect to happen when the ice melts?

SOLUTION

The water level will not change. Ice has a density far less than that of water, which is why it floats. The ice cubes will position themselves in the fluid based on the ratio of their density to that of the water. As the ice melts, it becomes water (more dense) and fills a smaller volume. The volume filled by the melted ice will equal the amount the water rose when the ice was first added.

A rule to remember when solving these conceptual problems is: **The ratio of the densities of an object to that of a fluid will equal the fraction of the volume submerged.** This idea makes sense because less dense objects should have less volume submerged and more of their volume floating.

in which ρ is the density of the fluid, **V** is the volume of the object or the fluid volume displaced, and **g** is the acceleration of gravity.

If an object is suspended from a string into a fluid (Figure 10-6), you can find the **tension** on the string by diagramming the forces involved. The upward forcesare tension and buoyant force. The downward force is weight. Tension will be (Weight) − (buoyant force).

II. Fluid Dynamics

A. CONTINUITY EQUATION

The **continuity equation** (Figure 10-7) says that the volume of fluid entering a pipe per unit time must equal the volume of fluid leaving the pipe per unit time, even if the diameter of the pipe changes:

$$\mathbf{A_1 v_1 = A_2 v_2}$$

EXAMPLE 10-2

AN ICEBERG FLOATING in the Arctic Sea has a density of 920 kg/m^3. Find the fraction of the iceberg that floats **above** the surface of the sea. Assume the density of sea water is 1025 kg/m^3.

SOLUTION

Find the ratio of the densities:

(Density iceberg)/(density seawater) = 920/1025 = 0.90

Thus, 90% of the iceberg is submerged and 10% of the iceberg floats above the sea surface.

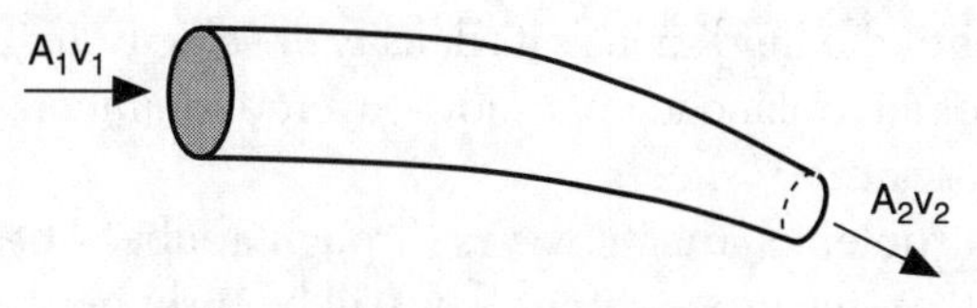

Figure 10-7. Continuity equation.

in which **A** is the cross-sectional area of the pipe and **v** is the velocity of fluid flow. This idea makes sense because what flows into a system must flow out. Because fluid is incompressible, the fluid flow rate through a large pipe and into a small pipe will increase in velocity. The product (A)(v) for the fluid will not change, however, which is the same as saying that the volume of fluid per unit time will not change.

Think about a garden hose. The flow of water from the end of the hose is the same whether or not you narrow the opening. When you do narrow the opening, you decrease the cross-sectional area of the hose, and the velocity of fluid flow increases.

B. BERNOULLI'S PRINCIPLE (Figure 10-8)

This important principle says that the quantity: $\mathbf{P + 1/2\rho v^2 + \rho gh = a\ constant}$ everywhere in a flow tube; **P** is pressure, **v** is velocity of fluid flow, and **h** is the height of the tube from a reference point.

Notice that the $1/2\rho v^2$ term resembles the kinetic energy expression, and describes the kinetic energy per unit volume of the fluid. The pgh term looks like the potential energy expression, and describes the potential energy per unit volume of the fluid. Thus, the **Bernoulli equation is simply a manipulated conservation of energy expression.**

The Bernoulli equation is useful in solving qualitative, conceptual questions. Suppose you are asked how the pressure will compare as fluid flows from a large rigid tube to a small rigid tube at the same height. To answer this question, realize that the velocity of fluid flow will increase from the large tube to the small tube (continuity equation).

$$\mathbf{P + 1/2\rho v^2 + \rho gh\ (large\ tube) = P + 1/2\rho v^2 + \rho gh\ (small\ tube)}$$

Because the KE term of the small tube is greater than that of the large tube (because of a higher velocity), and because the heights are the same, the pressure of the small tube must be smaller. Thus, the pressure of fluid flow at point B (Figure 10-8) is less than at point A. This concept may seem anti-intuitive, but remember you are solving a fluid dynamics problem, not a fluids-at-rest problem.

C. LAMINAR VERSUS TURBULENT FLOW

Laminar flow or streamline flow is best thought of conceptually. If fluid travels in cylindric "sheets" through a tube, each "sheet" slides by its neighbor. In this way, the fluid flow in a tube is directed in one direction without disorganized turbulence.

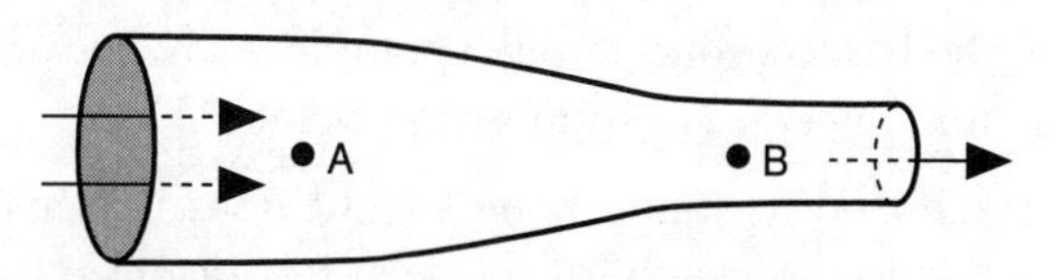

Figure 10-8. Bernoulli's principle. At point A, the value of the Bernoulli equation is equivalent to the value at point B.

When the properties of the fluid such as its density or velocity are increased, fluid flow tends to be **turbulent.** Consider an avalanche on a snow-covered mountain. Snow flows with disorganized turbulence, not in organized "sheets."

Fluid may flow in turbulent spiraling waves through a tube. **The Reynold's number** is a quantitative relationship that suggests whether flow will be turbulent or laminar/streamline. Values of the Reynold's number over 2000 suggest turbulent flow, whereas values less than 2000 suggest laminar flow.

$$\textbf{Reynold's number} = 2\rho vr/\eta$$

in which ρ is fluid density, **v** is velocity, **r** is radius of tube, and η is fluid viscosity.

D. VISCOSITY

Intuitively, viscosity depends on the "thickness" of a fluid. Syrup is considered viscous because it is "thick." To understand viscosity and how it is measured, consider **two flat plates separated by a thin fluid layer.** If the lower plate is fixed and a force is applied to the upper plate in an attempt to move it at constant speed, you could measure the force applied. **This force would be a measure of the viscous forces of the liquid. Viscosity is a measure of how fluid flows,** and is defined based on a relationship among an applied force to an area (flat plate), velocity (of upper plate), and separation (between plates).

$$\eta \text{ (viscosity)} = \frac{F/A}{v/l}$$

in which F is force applied, A is the area to which force is applied, v is velocity of the upper plate, and l is the distance between plates.

Conceptual Questions:

1. A man fills a "weightless" water balloon with water from a swimming pool. He ties a string to the end of the water balloon and lowers the balloon into the pool, suspended from the string, until it is completely submerged. What fraction of weight of the balloon (in air) must he exert to hold the balloon at a constant level in the pool?

   Solution: The man exerts no force. Because the fluid in the pool is the same as the fluid in the balloon, the weight of the water in the balloon equals the weight of the water displaced; i.e., the buoyancy of the surrounding water supports the water balloon.

   If the balloon contained a fluid that was twice as dense as the pool water, the man would have to pull up with a force equal to the apparent weight (weight submerged in the pool) of the balloon. To find the apparent weight of an object submerged, subtract the weight of the volume of pool water displaced from the weight in air of the fluid-filled balloon.

2. A fisherman sits on a pier and lowers a fishing line into the sea. On the end of the line is a spherical lead weight. When the weight is submerged halfway to the ocean floor, the fisherman estimates that he has to support only 5 pounds. As he lowers the weight still further, what happens to the force needed to support the weight?

   Solution: The force needed to support the weight does not change. The lead weight is buoyed by a force equal to the weight of the seawater displaced. This buoyant force does not depend on the depth submerged as long as the object is fully below the water surface and is not touching the ocean bottom.

3. Some swimming pools are constructed with concrete walls that are thicker at the bottom than toward the top because water pressure increases with depth. Pool A is 10×20 ft and 15 ft deep. Pool B is 30×50 ft and 10 ft deep. Which pool should be the stronger based on physics principles?

   Solution: The construction of Pool A, the deeper pool, should be the strongest. The walls of the pool are subjected to pressure that depends only on the depth of the pool and not on how much water it contains. Thus, the pressure on the walls of the pool is greater for the deeper pool.

Physics of Heat, Electricity, and Magnetism

emperature

Temperature and Heat

11

heat

The basics of thermal energy are defined and discussed in terms of thermal energy, temperature, thermal expansion, heat, heat capacity, specific heat, heat transfer, and conservation of energy. Chapter 12 addresses the topics of calorimetry and thermodynamics.

I. Temperature

A. THERMAL ENERGY

The thermal energy in a substance is the kinetic energy associated with the random motion of the atoms and/or molecules it contains. As the thermal energy status of a substance increases, so does the total kinetic energy associated with that substance. This kinetic energy allows for additional atom and/or molecule translation, rotation, and vibration.

B. TEMPERATURE

Temperature is an indicator of the average random kinetic energy of the atoms and/or molecules in a substance. Average random kinetic energy of substance atoms and/or molecules increases as temperature is increased. Although the term **thermal energy** corresponds to the total kinetic energy associated with a substance, **temperature** is a measure of the average thermal energy per atom and/or molecule.

Thermal energy is an **extensive property** (its value is proportional to the amount of substance present); temperature is an **intensive property** (does not depend on the amount of substance present). To make the distinction clear, compare a cup of tap water to a large lake, both with a temperature of 67°F. Although the two bodies of water have the same temperature (i.e., same average thermal energy per molecule), their thermal energies differ (i.e., total energy content).

Thermometers measure temperature. Classic thermometers measure temperature as a function of liquid mercury expansion and contraction. The details of thermometer design are of less importance, however, than a good understanding of **temperature scale.** A temperature scale is an arbitrary system used to quantify the temperature of a substance. The scale is constructed by assigning arbitrary values to points on the device associated with important thermodynamic events (usually the melting and freezing points of pure water at one atmosphere pressure) and dividing the spaces between these reference temperatures into equally spaced values.

The temperature associated with the freezing of water is the **freezing point,** and the temperature associated with the boiling of water is the **boiling point.** Unfortunately, the point (i.e., the value) on the temperature scale assigned to the melting or boiling of water varies according to which of three temperature scales is used.

The official (SI) unit of temperature is the **Kelvin,** abbreviated as K (not °K; no degree sign is used). The height of the liquid–mercury column at the time of water freezing is marked 273 K,

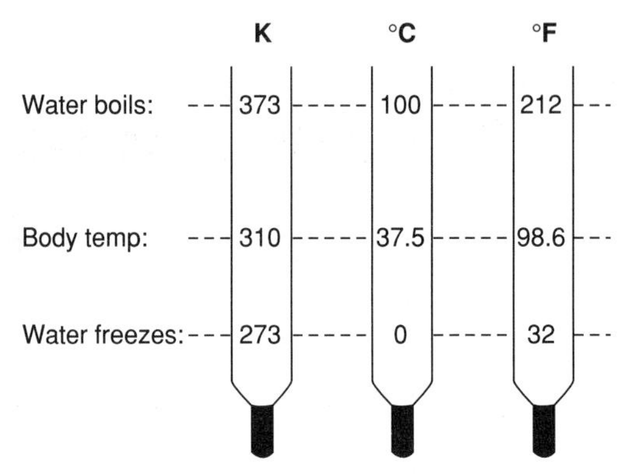

Figure 11-1. Three temperature scales.

and that associated with boiling is marked 373 K. Thus, a **temperature interval** (i.e., difference or delta) of 100 units exists between the two standard reference points.

The English system, or the "unofficial science language," defines the two reference points using the principle of "number simplicity;" the freezing point of water is zero degrees **Celsius** (0°C) and the boiling point is 100°C. Again, 100 units or degrees exist between the reference points of freezing and boiling.

Using the "American science language," the **Fahrenheit** temperature scale sets the freezing point at 32° Fahrenheit (32°F) and boiling at 212°F, creating a temperature interval between the two reference points of 180 units or degrees. Thus, it can be said that a one-unit temperature change in the Kelvin and Celsius systems is of a larger magnitude than one unit change in the Fahrenheit system. Another way of phrasing this same idea is that a change in one degree Fahrenheit (i.e., the American way) is "worth less" than a change in one degree of the systems used elsewhere. The key is that the differences in terminology are more than just semantics. The differences "between the languages" are the varying reference temperature values and intervals.

As shown in Figure 11-1, temperature scale conversions are straightforward. Computations to remember follow:

$$
\begin{aligned}
{}^{\circ}\mathrm{C} &= \mathrm{K} - 273 \\
T_F &= 9/5\, T_C + 32 \\
T_C &= 5/9(T_F - 32)
\end{aligned}
$$

Be aware that 0 K is a special temperature point and is referred to as **absolute zero** (i.e., the temperature at which all atoms and/or molecules possess zero kinetic energy).

Answer the following questions:

1. Can you calculate absolute zero in terms of °C and °F?
   Answer: −273°C and −460°F, respectively.
2. Is there a temperature less than 0 K?
   Answer: No. Absolute zero is the "absolutely" lowest temperature and energy state possible.

II. Thermal Expansion

Both solids and fluids can experience changes in dimensions (linear and volume) that correlate directly with the amount of temperature change. Most substances expand as they are warmed and contract as they cool. Thermal expansion and contraction are a direct result of altered (increased and reduced, respectively) atom and/or molecule movement within the material. Residents of regions with cool temperatures witness the linear changes in sidewalk and road cement (buckling) that occur during summer heat waves. Temperature-induced changes in both linear and volume dimensions occur.

The amount of change in dimension logically is proportional to the initial dimension of the object (length [L] or volume [V]), the amount of temperature change ($\Delta T = T_{final} - T_{initial}$), and some coefficient indicative of how easily that material changes dimension (α or β). Two equations express this principle:

Linear change equation: $\Delta L = \alpha L_{initial} (\Delta T)$
Volumetric change equation: $\Delta V = \beta V_{initial} (\Delta T)$

Alpha (α) is the **coefficient of linear expansion.** The units of this term must be reciprocal Celsius degrees ($°C^{-1}$) in order for a unit of length to result. Beta (β) is the **coefficient of volume expansion** for both solids and fluids. For a solid, $\beta = 3\alpha$ (length, width, and height are changing).

Conceptually, it makes sense that gasoline vapors expand more easily than rubber > aluminum > steel > concrete > glass. What would be a good way to remove an aluminum lid that is tightly covering a glass jar? Heat the lid slightly, allowing the aluminum to expand. The glass does not expand as much as the aluminum per unit temperature change.

III. Heat and Heat Transfer

A. HEAT

Heat is the amount of energy transferred from one object or body to another because of a difference in their temperatures. Heat (Q) is quantified in energy units of joules (J) according to SI conventions. Heat is commonly expressed in units of calories (cal), in which 1 calorie is the amount of heat required to increase the temperature of 1 g of water by 1 K or 1°C at room temperature (293 K). Note that **1 cal = 4.2 J.** When the term calorie is capitalized, it implies kilocalories. Thus, food companies and hospital dieticians use 5 Calories to mean what is really 5 kilocalories (5 kcal) or 5000 calories.

Heat capacity is the amount of heat energy needed to raise the temperature of a certain amount of substance by 1 K or 1°C. The amount of heat (Q) required to increase the temperature of a given substance from any initial temperature (T_i) to any final temperature (T_f) can be determined by knowing the heat equation:

$$Q \text{ (heat)} = m\, c\, \Delta T$$

in which m is the mass of the object, c is the specific heat capacity of that particular substance, and ΔT is the change in temperature.

The **specific heat capacity** of a substance is the heat capacity per unit mass (i.e., the amount of heat that must be added to 1 g of that substance to raise its temperature by 1 K or 1°C). **Specific heat is the amount of heat in *calories* required to raise the temperature of 1 g of a substance by 1°C.** It is important to know specific heat values of water and ice (1.0 and 0.5 cal/g°C, respectively). Sample problems are provided in Chapter 12.

B. HEAT TRANSFER MECHANISMS

Heat is transferred from one region or body to another by the following basic mechanisms.

Convection involves the actual **movement** of part of the system from an area of high temperature to one of lower temperature. This process occurs in both gases and liquids. **Examples** of this process include movement of hot air pockets in a gas oven or in the sky.

Conduction involves thermal energy **transfer** without the macroscopic movement of the medium that carries the heat. Thus, like sound and electricity, heat conduction requires contact between neighboring molecules and molecule-to-molecule transfer of energy. Heat of conduction varies directly with surface area of the heat source and difference in temperature between the source and recipient locales. Heat of conduction varies inversely with distance from the source (i.e., the farther you are from the heat source, the less heat is transmitted to you). **Examples** include transfer of heat across a hot metal grill, around a cooking pan, or up the handle of a soup spoon.

Radiation involves the transfer of heat by **electromagnetic (EM) radiation.** Un-like convection and conduction, radiation can transmit heat energy through empty space. All forms of EM radiation have energy associated with them and, thus, heat energy. **Examples** include microwaves, ultraviolet light, x-rays, and deadly gamma rays. Remember: heat of radiation is proportional to the area (A) of the emitting surface, several constants (related to how good a radiator a particular object is), and to the fourth power of temperature (T) in units of Kelvin; heat of radiation is proportional to T^4 (where T is in units of K).

Question: What is the difference in the heat of radiation of Object A with a temperature of 1 K and Object B with a temperature of 2 K? **Answer:** Object B emits about 16 times more radiation than Object A if all other values are equal ($2^4 = 16$). Remember also that all objects above absolute zero (0K) emit radiation.

IV. Conservation of Energy

Energy is a **conserved** property (i.e., the total energy of a closed system is **constant**). Energy can be transferred between objects within a sealed system and/or can be converted to different forms. It makes sense that thermal energy (TE) change is one component of the energy conservation equation because energy in a system can exist as or be converted to thermal energy. Heat (Q) is also part of the energy conservation equation, because heat is a source of energy input into a system. If the total energy of a system is constant, then an expression that accounts for energy dynamics follows:

Heat + Work (input/output) =
Sum of changes in different forms of energy (KE, PE, and TE)

The **conservation of energy principle** states that what energy goes into or comes out of a system in the form of heat or work leads to a change in the various energy forms within the system (such as changes in kinetic energy, potential energy, and/or thermal energy).

Thermodynamics 12

The goal of this review of thermodynamic principles, the science of heat and temperature, and the laws governing the conversion of heat into other forms of energy, is to ensure strong qualitative and quantitative understanding of thermodynamics.

I. Thermodynamic Principles

Before solving thermodynamic problems, it is important to understand certain terms. Figure 12-1 displays a sealed glass box that contains a solid metal ball. Given that the box and ball are our main interests, we call the inside of the box the **"system"** under study. The system therefore comprises: (1) the air inside the box, and (2) the metal ball itself. The area outside of the system, or outside the box in this example, is the **"surroundings."**

If the system is **"closed"** (i.e., sealed), the system does NOT interact or exchange energy with the surroundings. If no energy in the form of heat can enter or leave the system, the net change in the system's overall energy and heat content is zero. If the system is **"open,"** the system and surroundings are free to interact and exchange energy forms. Systems can be open or closed.

Thermodynamic equilibrium exists if the measurable physical parameters of a system (e.g., temperature, volume, and pressure) are constant over time. **Thermal equilibrium** exists if two systems are in thermal contact and no heat flow between them occurs (their temperatures are equal).

If no change or net change in heat content occurs, then **$Q = 0$.** If a system gains energy in the form of heat, then the heat content increases [i.e., **Q is positive** (+)]. If a system releases heat, **Q is negative (−).** Recall the relationship: $Q = m\,c\,\Delta T$; Q is positive (+) if $T_{final} > T_{initial}$ and negative (−) if $T_f < T_i$.

The principle of **phase change** also warrants review. Many substances have the potential to exist in three states or phases: solid, liquid, and gas. The phase of a substance at any time (t) is related solely to its energy state at that particular time.

Figure 12-2 shows an experiment that involves the study of the phase changes experienced by water (H_2O) as it is heated. The H_2O used is pure (distilled) and is placed in a room at one atmosphere pressure. A hot Bunsen burner is placed below a 5-g block of ice (solid H_2O) with an initial temperature of −20°C (point **A** on the graph in Figure 12-2). As heat is supplied to the ice cube, the average random kinetic energy (KE) of the H_2O molecules increases, and they move more and more.

The ice is warmed until it reaches 0°C (point **B**), the **melting point** for pure H_2O at 1 atm pressure. Note that although the temperature of the entire ice cube is increasing, no change in phase has occurred (i.e., it does not melt at all). Not until **all** regions of the ice cube (both the inside and outside) reach the melting point (0°C) does a change in phase occur (i.e., melting of the

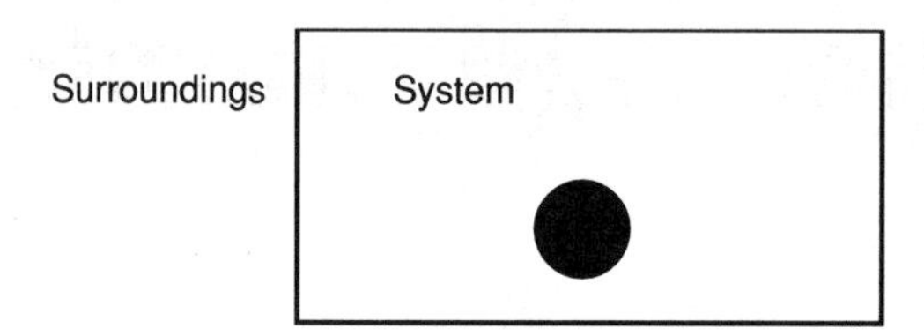

Figure 12-1. A system and surroundings.

ice cube begins). Then, energy supplied by the hot flame goes to melt the ice cube only, with no ΔT during the melting process. Not until every molecule changes phase (solid → liquid) does the temperature again begin to increase.

Point C represents newly formed water (the liquid form of H_2O) at 0°C. Heat from the flame then warms the water. The phase of the water does not change at all until every region of the flowing fluid reaches 100°C (point **D**), the official **boiling point** for pure water at 1 atm pressure. As heat is absorbed by the water, the energy level of the water molecules increases, leading to **vaporization** (molecules escaping the surface of the liquid because of their great energy level).

Figure 12-2 shows that temperature always remains constant during a phase change. Water at 100°C is converted to vapor at 100°C. As long as even one water molecule at 100°C is still around, the temperature of the steam will not rise above 100°C. Once all the liquid (i.e., water) at 100°C has been converted to gas (i.e., water vapor) at 100°C (point **E**), the temperature of the gas begins to rise.

The phase change from B → C is **melting** or **fusion;** the change from D → E is **evaporation.** If a hot object releases energy in the form of heat, it can move from E → D in a process termed **condensation.** If more energy is released from the object, it can move from C → B, which is termed **freezing.** The process involving the direct conversion of either the solid to the gas form (B → E) **or** gas to solid form (E → B), without going through the liquid phase, is **sublimation.** Gases, when exposed to extremely high energy, may decompose into hot charged particles referred to as **plasma** (occurs at point **F** on Figure 12-2; p, plasma phase of matter).

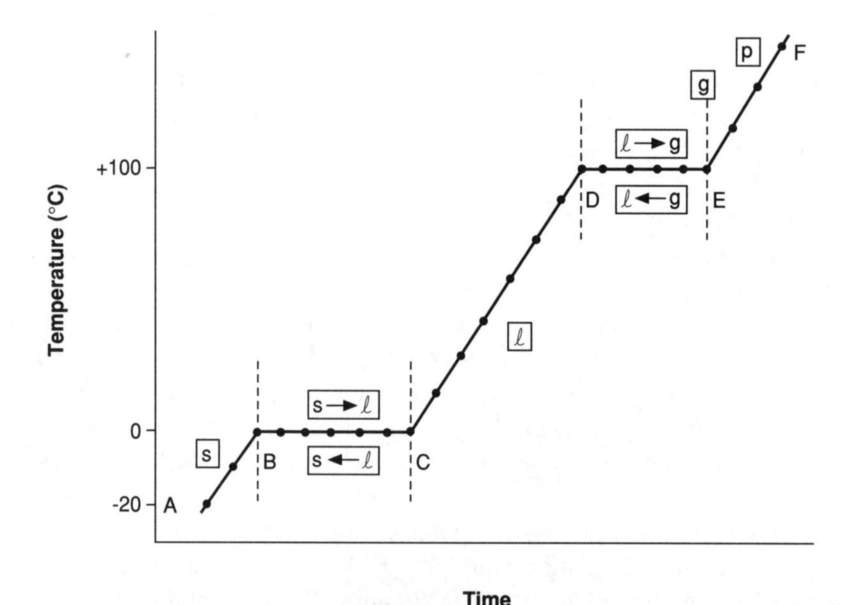

Figure 12-2. Phase changes of water (H_2O). s = solid, l = liquid, g = gas, p = plasma.

Physicists define the **heat of fusion (H_f)** of a substance as the amount of heat needed to transform 1 g of a substance from a solid to a liquid. The H_f of ice is 80 cal/g. The **heat of vaporization (H_v)** is the amount of heat required to convert 1 g of a liquid into a gas. The H_v for water is 540 cal/g. Clearly, it takes more energy per gram sub-stance to vaporize than to melt (almost seven times as much energy per gram). Similarly, 1 g of steam undergoing condensation releases more energy than 1 g of water under-going the freezing process (also about seven times more energy per gram given off). The heats associated with the two phase changes just mentioned can be calculated for any substance given the H_f and H_v values and the mass (m) of substance going through the phase change:

$$Q_{fusion} = mH_f \qquad Q_{vaporization} = mH_v$$

("interphase" heat equations)

In this chapter, the term **"interphase"** refers to the heat involved in the phase change occurring between two different phases. The preceding formulas are used **only** for "interphase" processes. To calculate the amount of heat needed to heat water from 20° to 30°C, or the amount of heat released when cooling water from say 30° to 20°C, use the **"intraphase"** (i.e., within a single phase) equation:

$$Q = m\,c\,\Delta T \qquad \text{("intraphase" heat equation)}$$

The specific heat capacity (c) of ice is 0.5 cal/g°C, whereas that of water is 1.0 cal/g°C. This equation is used to determine how much heat is required or released from $T_{initial}$ to T_{final}. Remember that ΔT is defined as the change in temperature ($T_{final} - T_{initial}$). (See the sample problems illustrating the use of these expressions in the Calorimetry section.)

II. Calorimetry

Calorimetry is the science of "counting calories." Calorimetry is often used when quantitating energy gained or lost during some chemical reaction, biologic process, or activity. Calorimetric techniques are based on principles of energy (thermal energy) conservation, which were discussed previously.

The **calorimeter** is a well-insulated device that allows for the creation of a closed (sealed) thermal system. Ideally, no transfer of heat should occur from the surroundings into the system or from the system into the surroundings. Such ideal conditions are realized by excellent calorimeter design and construction. The calorimeter consists of a durable metal, glass, or other strong container that usually is surrounded by a good insulator, such as styrofoam, which prevents heat transfer. The container is filled with a material of known composition, temperature, and thermal behavior. The test object, with an unknown thermal content, is then introduced into the container. Such a calorimetric setup is demonstrated in Figure 12-3.

If a researcher wants to know the amount of heat released by a chemical reaction, biologic process, or mass, he inserts the test object into the container. It is assumed that any change in the temperature and/or state of the fluid is related to the energy content of the test object. It is now possible to quantitate thermal processes, given that the heat energy gained by the colder object must equal that lost by the warmer object (i.e., **no net change in the heat content of the system**). These statements are equivalent but can be expressed separately as equations:

$$Q_{gained} = Q_{lost} \quad \text{or} \quad \Delta Q = 0$$

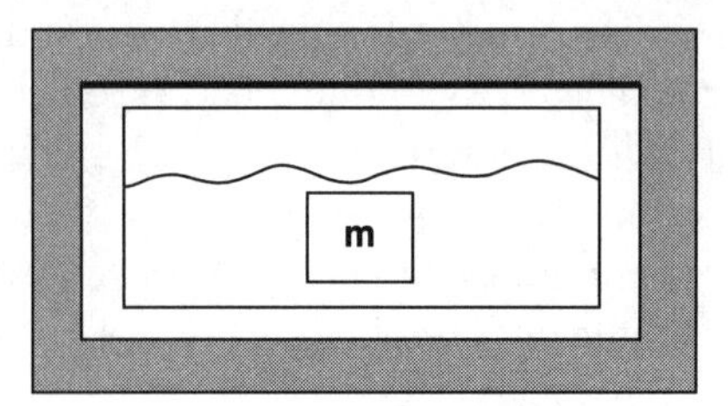

Figure 12-3. A calorimeter.

which is equivalent to:

$$\mathbf{mc\ \Delta T_{gained} = mc\ \Delta T_{lost} \quad or \quad (mc\ \Delta T_{gained}) + (mc\ \Delta T_{lost}) = 0}$$

Calorimetry can be used to determine any of the following variables found in the heat content equation if all others are known: final temperature of a mixture, specific heat capacity of some unknown material, and the mass of a material present in the container.

Note that each quantity on both the left and the right are positive; you are summing each side, assuming total heat acceptance or contribution.

EXAMPLE 12-1

AS SHOWN IN FIGURE 12-3, a 25-g block of pure metal (unknown type) with a temperature of 100°C is placed into a well-insulated container filled with 50 g of water ($c = 1.0$ cal/g°C) at 20°C. The lid of the container is tightened to ensure total insulation. The metal is completely submerged in the water and the mixture is allowed to sit for 5 minutes. The thermometer that penetrates the lid shows the final temperature for the water bath is 24°C. What is the specific heat capacity of the metal?

SOLUTION

Taking into consideration that a "small hot" mass is immersed in a "moderate" volume of "cool" water, it is reasonable to expect the water to warm slightly as a result of thermal energy (heat) released by the hot metal. Determine c_{metal} using the principles discussed previously.

Using the formula $Q_{gain} = Q_{lost}$ is often a faster and safer way to calculate unknowns in calorimetry problems because you need not assign positive or negative signs to Q values. Be sure, however, that each quantity in the expression has a positive value (because you are summing values to be used for equivalency).

When final temperature is less than initial temperature, as is in the following example (*), the normal definition of ΔT ($T_f - T_i$) is reversed (changed to$T_i - T_f$) so that the quantity itself is positive. Because a phase change in this experiment is unlikely (given the numbers provided), it is possible to use theintraphase heat transfer expression: $Q = m\ c\ \Delta T$.

$$\begin{aligned} Q_{\text{gained by the water}} &= Q_{\text{lost by the metal block}} \\ (m)(c)(T_f - T_i) &= (m)(c)(T_i - T_f)^* \\ (50\text{ g})(1.0\text{ cal/g°C})(24 - 20°\text{C}) &= (25\text{ g})(c)(100 - 24°\text{C}) \\ 200\text{ cal}/(25\text{ g} \times 76°\text{C}) &= 0.1\text{ cal/g°C} = c\text{, answer} \end{aligned}$$

Example 12-2

FIFTY GRAMS OF ICE at $-2°C$ is added to 10 g of steam at 100°C in a sealed chamber. What is the final temperature of the mixture if no heat can enter or leave the system?

SOLUTION

Keep in mind that the steam gives up heat energy to the ice, and the ice gains heat energy from the steam. The system reaches an equilibrium point (i.e., only one phase). Also, energy gained by the ice is equal to that lost by the steam. Start by predicting what happens to each substance as the experiment continues. You have 50 g of "slightly cold" ice (which needs only slight warming to reach a phase change location) and only 10 g of "very cold" steam (100°C is the lowest temperature at which pure steam can exist before cooling [condensing] to become liquid [water]).

Q_{gain}:

i. Ice probably warms to 0°C, and then:
ii. Ice probably goes through a phase change (melting) to become water at 0°C, and then:
iii. Water at 0°C probably warms to some T_f that corresponds to the liquid temperature range.

Q_{lost}:

iv. Steam already at 100°C probably goes through a phase change (condensation) to become water at 100°C, and then:
v. Newly formed hot water at 100°C probably cools to some T_f in the liquid temperature range.

To set up the quantitative expression and solve for T_f:

$$\text{Heat gained by ice} = \text{Heat lost by steam}$$

$$(mc\Delta T)_{ice} + (mH_f)_{ice\rightarrow water} + (mc\Delta T)_{water} = (mH_v)_{steam\rightarrow water} + (mc\Delta T)_{water}$$

$$(50\text{ g})(0.5\text{ cal/g°C})(0-[-2]) + (50\text{ g})(80\text{ cal/g}) + (50\text{ g})(1\text{ cal/g°C})(T_f - 0)$$

$$= (10\text{ g})(540\text{ cal/g}) + (10\text{ g})(1)(100-T_f)^*$$

$$50\text{ cal} + 4000\text{ cal} + 50\mathbf{T_f} = 5400\text{ cal} + (1000 - 10T_f)$$

$$60\mathbf{T_f} = 2350$$

$$\mathbf{T_f} = 39°C \text{ (liquid phase)}$$

Calorimetry techniques can be used to determine basal metabolic rate. A person is placed in a small, closed chamber filled with cool air of known mass (m), specific heat (c), and temperature (T). Energy in the form of heat leaves the warm person and heats the air circulating through the chamber. The amount of energy per unit time released from the subject is calculated by using equipment outside the chamber.

III. Work

The discussion of work in Chapter 5 dealt with mechanical work. The following section also addresses mechanical work but with the additional consideration of pressure–volume systems acting as energy sources.

A metal container has a piston-type lid that moves up and down depending on the pressure underneath it (Figure 12-4). The container is filled with gas molecules, which when heated, tend

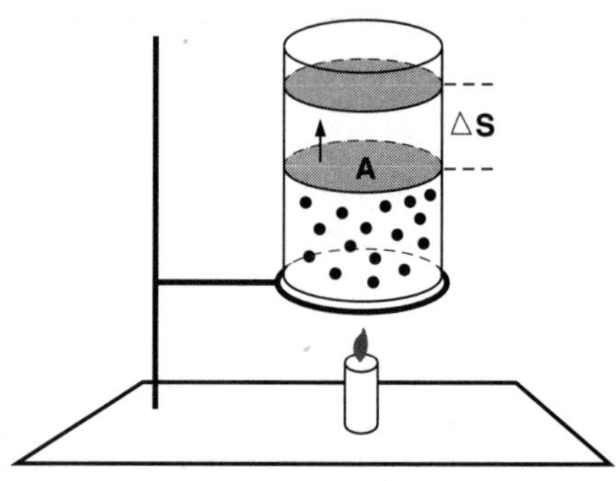

Figure 12-4. A Bunsen burner heating a piston system. As the kinetic energy of the gas molecules increases, pressure increases in the piston system, which leads to a volume increase.

to bump harder against the walls and lid of the container. As the gas molecules obtain greater and greater kinetic energy from heating, the pressure inside the container increases.

If the lid, with surface area A, rises a distance Δs, then the work done on the system(i.e., the container) by the surroundings (the Bunsen burner) can be derived by using the following concept: work is proportional to the force applied, the distance over which it is applied, and the angle between force and direction of travel [considered to be 0°, so ø = 1].

$$\mathbf{W = (F)(\Delta s) = (P \bullet \mathit{A})(\mathit{\Delta s}) = (P)(\mathit{\Delta V})}$$

Thermodynamic work is equal to the pressure of a system times the change in volume produced as a result of a force. This simple expression assumes constant pressure conditions throughout the work process (more of a rarity in real life).

According to common physics conventions, work done **by the system** is considered "positive" (+) work, whereas work done **by the surroundings** is "negative" (−) work. This designation is arbitrary and may disagree with the conventions used in college chemistry and physics courses.

The logic behind the signs is better appreciated by knowing the corresponding internal energy expression **(first law of thermodynamics),** which states that the change in the internal energy status of the system (ΔU or ΔE) is equal to the change in heat content of the system (ΔQ) **minus** the work done by the system:

$$\mathbf{\Delta U \text{ (or } \Delta E) = \Delta Q - W \text{ (First law of thermodynamics)}}$$

It follows that if the system does work (W+), the overall internal energy status of the system is less (i.e., ΔU or ΔE is < 0) (see subsequent discussion in this chapter).

The PV work done by or on the system can be illustrated graphically. If pressure is plotted on the ordinate (y or vertical axis) and volume on the abscissa (x or horizontal axis) for several different times, a PV curve is generated (Figure 12-5). In this experiment, a piston is locked to compress the gas molecules residing within the chamber leading to a relatively low volume and high pressure state (point A on Figure 12-5). The process leading to a smaller volume and higher pressure is a **compression.**

The piston is allowed to move freely and the volume of the chamber expands, allowing the gas molecules to occupy more volume of the chamber and leading to lower total pressure (point B on Figure 12-5). The work associated with this **expansion** is positive and can be determined by measuring the area underneath the experimental PV curve.

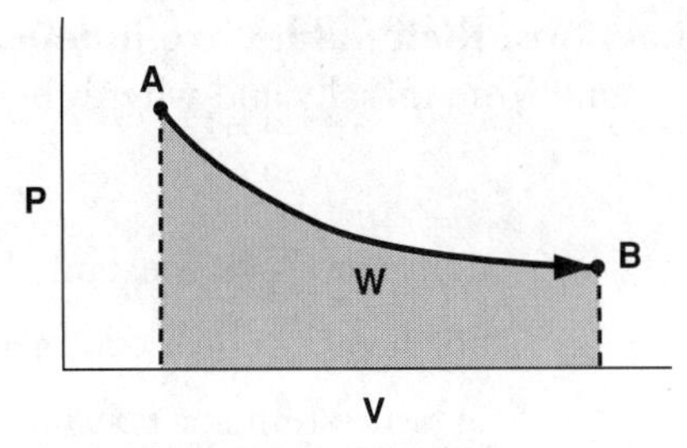

Figure 12-5. A representative PV diagram.

The area under a complex PV curve can be difficult to approximate. Because the work definition holds true if P is constant, however, only those situations in which compression (a decrease in volume) or expansion (an increase in volume) occur with no or minimal change in system pressure require such an approximation.

A more typical PV diagram is shown in Figure 12-6. After studying this curve, answer the following questions: What process occurs as the piston moves from Point A to Point B? What are the units of work? How much work was done? Assume 1 atm $\approx 1 \times 10^5$ Pa.

Solution: The process occurring from points A to B is an **isobaric** (in Latin, *iso* means same, *baric* refers to pressure) **expansion**. Units of work in the SI system are Pascals (pressure unit) and m^3 (volume unit). The units of Pa and m^3 multiply to yield joules. Because the graph in Figure 12-6 shows a pressure of 2 atm, the pressure in Pascals is $(2)(1 \times 10^5) = 2 \times 10^5$ Pa. The volume change is $(2.0 - 1.5) = 0.5\ m^3$. Thus, $W = P\ \Delta V = (2 \times 10^5\ Pa)(0.5\ m^3)$ $= \mathbf{1 \times 10^5\ J}$.

IV. Laws of Thermodynamics

A. FIRST LAW OF THERMODYNAMICS

$$\Delta U\ (\text{or}\ \Delta E) = \Delta Q - W$$

The first law states that the **energy of an isolated system is constant.** The internal energy status of the system (ΔU or ΔE) varies with changes in heat content (ΔQ) and work status (W) of the system. The internal energy status of an isolated system decreases(ΔU −) if heat is lost and/or if the system does work on the surroundings. The internal energy status increases (ΔU +) if the system gains heat and/or had work done to it by the surroundings.

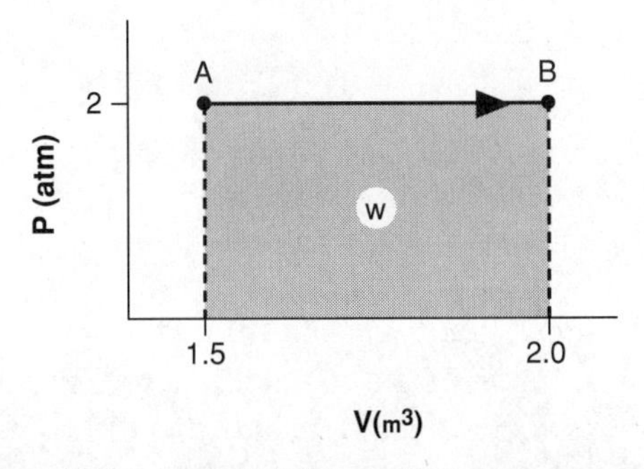

Figure 12-6. Typical PV diagram. Note constant pressure.

ΔU and ΔQ are state functions; their values are independent of specific pathway. Important facts are where they came from initially and where they are now. Common thermodynamic terms include:

Isothermal:	temperature is constant, $\Delta U = 0$
Adiabatic:	no change in heat content, $\Delta Q = 0$
Isobaric:	pressure remains constant, $\Delta P = 0$
Isovolumetric:	volume remains constant and so $W = 0$

In an isothermal process, the change in heat content equals the work done. In an adiabatic process, the change in internal energy status equals work done. In an isovolumetric process, the change in internal energy equals the change in heat content.

B. SECOND LAW OF THERMODYNAMICS

$$\Delta S \geq 0$$

The second law states that **entropy (disorder) always tends to increase.** Consider the entropy of a gas > liquid > solid, or disorder in everyday life that never seems to decrease. Entropy is abbreviated S; **entropy change** (change in amount of disorder) is abbreviated ΔS.

The second law allows for the following processes:

1. $\Delta S > 0$, which represents the spontaneous and irreversible processes that happen in nature: (e.g., balls initially at rest rolling down a steep hill toward the center of gravity; a clump of ants dispersing throughout the environment; heat flow from a hotter to a cooler object). In these cases, ΔS is positive (+).
2. $\Delta S = 0$, which states that disorder is not changing now, but it will change soon. Such processes are reversible, because they can become spontaneous and irreversible at any moment: (e.g., a ball at rest on a mountain peak that stays at rest until "a natural phenomenon," such as a wind gust or earthquake gets the ball rolling).

A relationship exists between heat content (Q), entropy (S), and free energy of a system (G): $\Delta G = \Delta Q - T\,\Delta S$. If a spontaneous process (ΔS +) occurs in a sealed system ($\Delta Q = 0$), then the free energy of the system (ΔG) must be negative (−). Therefore, ΔG is negative and ΔS is positive if a process is spontaneous. (For more discussion of this topic, see *High-Yield General Chemistry*)

Electrostatics 13

The focus of this discussion is on fundamental concepts relating to electrically charged particles at rest, including the definition of "electrical charge" and how charges interact with each other and their environment. Subsequent chapters address the topic of moving charged particles, assuming familiarity with the basic physics of "stationary charges."

By learning and thinking about electrostatics in a new, simplistic way, reviewing basic mechanical principles in preceding chapters, and tying concepts together, you will gain a solid understanding, both conceptually and quantitatively, of static and dynamic electrical phenomena.

I. Charge and Related Topics

A. CHARGE

Electric **charge** can be a property of a particle, substance, object, or body. A net electric charge exists if, and only if, there is a net inequality in the number of **electrons** and **protons** in a given atom. Remember that the neutron, a third subatomic particle, lacks charge. Also, whereas the mass of an electron is only 1/2000 that of a proton, the magnitude of the charge on each is the same, 1.6×10^{-19} coulombs (SI unit of charge is the coulomb [C], the experimentally determined net charge of 6×10^{18} protons). The sign of the charge for the electron is negative (−), whereas that of the proton is positive (+). Thus, $|e-| = |p+| = e = 1.6 \times 10^{-19}$ C. As an illustration, the spark seen when touching a doorknob is about 50 nC, whereas the average lightning bolt discharges approximately 5 C. The most common letter abbreviation for the magnitude of a charge is **Q** or **q** (e.g., $Q = 2 \times 10^{-7}$ C).

Electrons and protons are the fundamental "players" in the charge game. An atom, by definition, has no net charge (is electrically neutral overall) because it has equal numbers of electrons and protons. Although materials with an equal number of electrons and protons are electrically **"neutral"** overall, regions of charge imbalance may exist. Materials with an unequal number of electrons and protons are said to be **"charged."** Figure 13-1 shows a positively charged object.

Some atoms easily gain or lose an electron when interacting with other atoms. An atom that has lost an electron has a net positive charge (+e) and is called a **positive ion.** An atom that gains an electron has a net negative (−e) charge and is called a **negative ion.** Regions of a substance, object, and/or body can become charged by accumulating charged particles while remaining electrically neutral as a whole. Examples include common daily materials and tools, rocks, clouds, the human heart and nervous system, and the like.

B. CONDUCTORS, SEMICONDUCTORS, AND INSULATORS

Metal is a material that can accept and/or lose electrons easily and become electrically charged. Metals are generally referred to as **conductors** because they often are able to conduct electrical

Figure 13-1. A charged object.

charge along their length. They may pass electrons from neighbor to neighbor, region to region, and end to end. Other good conductors include salt solutions, ionic compounds, and certain gases.

Insulators tend to prevent the flow of charge (electrons). Common electrical insulators include glass, rubber, wood, ceramic, and plastic. **Semiconductors** are materials that have a variable ability to conduct charge. The ability of a semiconductor to conduct charge varies with ambient temperature, becoming a better conductor (more electrons available) at higher temperatures and a poorer conductor (less free electrons) at lower temperatures. Frequently used semiconducting materials include silicon, carbon, and germanium.

Materials that are good thermal (heat) conductors tend to be good electrical conductors, whereas materials that are poor thermal conductors tend to be poor electrical conductors.

II. Coulomb's Law and Electric Force

If you place a small, positively charged metal ball (charge $+Q_1$) a short distance (r) to the left of a second positively charged metal ball (charge $+Q_2$) at rest (Figure 13-2), you would expect, based on experience with charged materials (e.g., magnets), that the two charges will move away from each other. It makes sense that the repulsion action allows for the production of an **"electric force"** that tends to push outward laterally from each ball (shown as arrows). This scenario has overtones relating to Newton's law of gravitation (see Chapter 7). The amount of force exerted on a ball is directly propor-tional to the magnitude of the charges involved (Q_1 and Q_2), inversely proportional to the square of the distance (r) separating the charges involved, and proportional to some constant (k). This deduction is the basic expression experimentally determined by Charles Coulomb in 1788:

Coulomb's law: $$\mathbf{F}_{(\text{of Q1 on Q2})} = \frac{\mathbf{k}|\mathbf{Q_1}||\mathbf{Q_2}|}{\mathbf{r}^2}$$

Coulomb's law determines the magnitude of electrostatic force between two charges at rest. Recall that units of force, charge, and distance are Newtons, coulombs, and meters, respectively. Can you deduce the proper units of k? (**Answer:** Nm^2/C^2.)

As reviewed in Chapter 1, a force is a **vector** quantity, and thus, has both a magnitude and direction. Assuming the charges are static, it is possible to calculate the **magnitude** of the (electric) force that charge 1 places on charge 2, and vice versa. To recall the direction of any electric

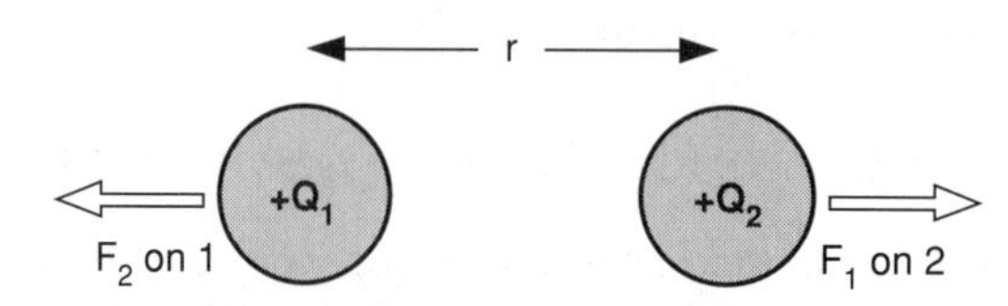

Figure 13-2. Forces associated with two positively charged objects.

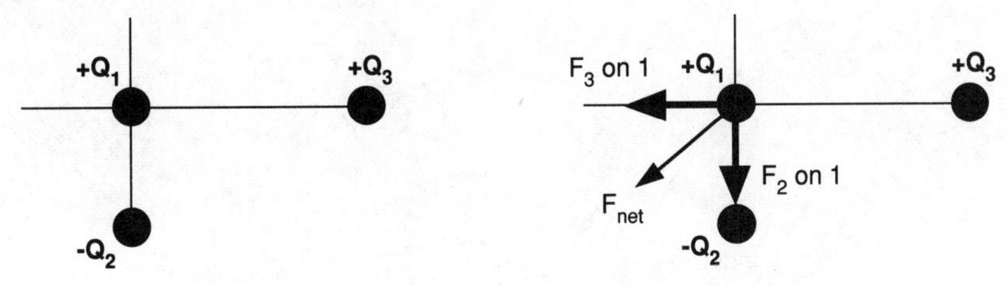

Figure 13-3. Determining net forces involving multiple charges.

force, remember the saying: "likes repel, opposites attract." Similarly charged objects repel (move away from) one another, whereas oppositely charged objects attract (move toward) one another.

Question 13-1: Can you state the similarities and differences between gravitational and electrostatic forces?

Answer: Both of these forces are "real" forces between particles, vector quantities with both magnitude and direction, and inverse square laws; they obey Newton's third law, act through space, and must be added vectorially. These forces differ in that gravitational forces are "attractive" only and usually are weak, whereas electrical forces are attractive or repulsive and are strong (about $10^{40} \times F_{grav}$ on an atomic scale).

Figure 13-3 illustrates a more complicated problem involving **multiple** neighboring charges. In this coordinate system, Q_1 (+2 mC) is located 2 m to the left of Q_3 (+3 mC) and 1 m above Q_2 (−1 mC). Note that all three charges are within the same plane.

Question 13-2: What is the net electric force experienced by Q_1 if $k = 9 \times 10^9$ Nm^2/C^2? (**Remember:** "force" means both magnitude and direction relative to some reference point; also, mC means 10^{-6} C.)

Solution: Calculate the magnitude and direction of the net electric forces acting on charge 1. Using Coulomb's law, find that the force (F) of charge 3 on charge 1 equals $(k)(Q_3)(Q_1)/(r^2)$, which equals 0.014 N. The force (F) of charge 2 on charge 1 equals $(k)(Q_2)(Q_1)/(r_2)$, which equals 0.018 N.

Using techniques of vector summation, the resultant force $(F_R) = [(0.014\ N)^2 + (0.018\ N)^2]^{1/2} =$ **0.023 N.**

Because the direction of the electric force of charge 3 on charge 1 is toward the $-x$ direction and that of charge 2 is in the $-y$ direction, and because the magnitudes of the two forces are similar, the net direction is about midway between the negative horizontal and vertical axes. The exact angle below the horizontal can be determined by : $\tan \theta = (F_y/F_x)$ and thus, $\theta = \arctan (F_y/F_x) = 52°$ below the $-x$-axis.

Question 13-3: What is the net electric force on a +5 mC point charge if it is surrounded symmetrically at a distance of 0.8 m by a metal hoop, the entire surface of which is covered with a charge −15 mC?

Answer: Because the sole positive charge is attracted to all points on the surrounding ring equally, the net electric force is **zero**—all points on the ring "enjoy" the presence of the positive charge equally.

III. Electric Dipole

In Figure 13-4, rod A is charged with a positive charge, and rod B is charged with electrons by rubbing rabbit fur across it several times. Two simple experiments are performed. In the first, as

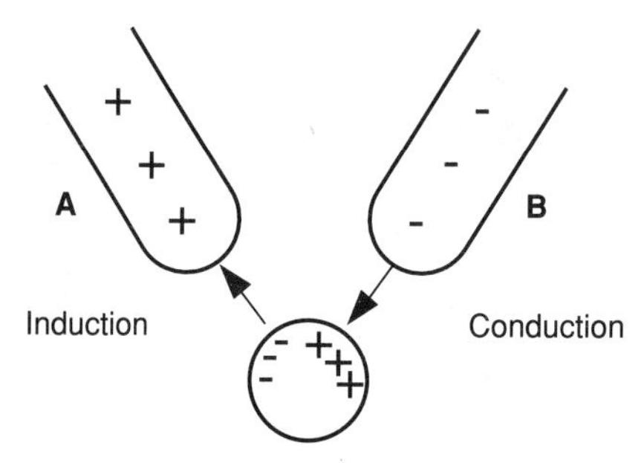

Figure 13-4. Conduction giving rise to induction of a dipole.

rod B moves toward a round metal ball, the positive charges within the metal ball move toward rod B because of electrical attraction. By lightly touching the rod to the ball (i.e., contact), **conduction** of electrical charge occurs. If in a second experiment, however, rod A is kept a slight distance from the metal ball, a special situation is created—**induction** of a dipole.

An electric dipole is induced by the separation of oppositely charged particles (or regions) of equal magnitude—+Q and −Q are separated by a small distance (r). The electric dipole moment is a **vector** quantity. The magnitude of the electric dipole is expressed as follows:

$$\textbf{Electric dipole moment} = \mathbf{p} = |\mathbf{Q}|\,\mathbf{r}$$

The direction (assigned by most physicists) is **from negative to positive charge** (i.e., [−] → [+]). As detailed in the next section, this direction is exactly opposite to that of electric fields (E-fields), which, by convention, are directed from positive to negative charge (i.e., [+] → [−]). Unfortunately, the direction of the dipole vector varies by specialty; chemists usually state that direction of a dipole is from the positively charged center of one atom to the negatively charged center of another. By classical physics conventions, the dipole moment is oriented from negative to positive charge (a dipole tends to align **with** the field).

IV. Electric Fields

A small positive test charge (+q) is placed near a negatively charged metal ball (−Q) (Figure 13-5). What net effect will the metal ball have on the test charge?

The +q will be attracted toward the metal ball, and thus experience a force (i.e., electric force) to the right. If you plot the relative force directions for a series of positive test charges (+q), you display **lines of force.** In actuality, lines of force are graphic representations of the **E-fields** produced around the positively charged metal ball. All charged particles, objects, and bodies set up an E-field around them. The density of lines drawn represents the relative magnitude of the E-field. The more lines per unit area, the greater the E-field strength. The line direction shows E-field direction.

To determine the lines of force (E-field) around any charge(s), determine the direction of the force experienced by a series of small test charges (+q) when placed in

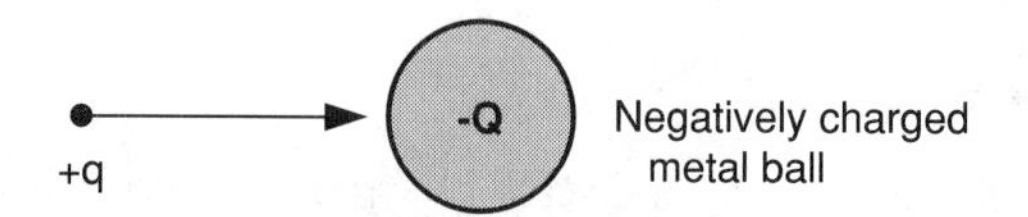

Figure 13-5. Small positive test charge placed in the region of a negatively charged metal ball.

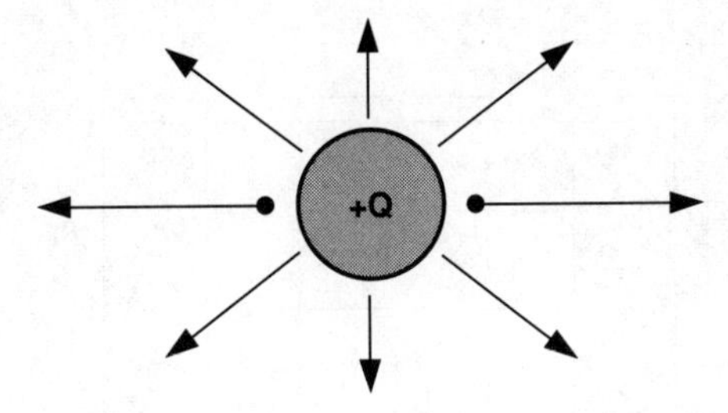

Figure 13-6. Effect of a charge of interest on a positive test charge (+q).

the vicinity of the charge(s) of interest. Figure 13-6 shows how test charges leave the region of +Q of a positively charged metal ball and head into the periphery.

Consider the situation of two equal magnitude charges positioned side by side. Figure 13-7 illustrates the E-fields around two oppositely charged objects (at right) and two positively charged objects (at left). As the E-field lines show, if you place a positive test charge in the vicinity of the oppositely charged objects, it is attracted to the negatively charged object and repelled by the positive. On the other hand, the positive test charge is repelled from both positively charged objects.

Note how the positive test charges minimize contact with positive charges while maximizing contact with negative charges. All charged objects in the world are surrounded by these amazing "invisible" E-fields.

E-fields are vector quantities with a definite magnitude and direction. They are defined in terms of the force acting on a small, positive test charge placed in the field rather than in terms of the charges causing the field. The E-field, however, is solely determined by the charges causing the field, not the test charges. Thus, we use the concept of the test charge to find out what we want to know (i.e., the magnitude of an unknown E-field). The magnitude of a given E-field is calculated as follows:

$$\mathbf{E} = \frac{\mathbf{F}}{\mathbf{+q}} = \frac{\mathbf{kQ}}{\mathbf{r^2}}$$

The E-field at a point in space is the ratio of the net electric force (F) acting on a small, positive test charge (+q) placed at that point, divided by the magnitude of that test charge.

Because $\mathbf{F} = kQq/r^2$, and $\mathbf{E} = \mathbf{F}/q$, then $\mathbf{E} = kQ/r^2$, which gives the magnitude of the E-field at a known distance (r) from a charge (Q). Be careful not to confuse this calculation with the expression for electrical potential (V = kQ/r; see Chapter 14).

The unit of E-field magnitude is N/C. The direction of the E-field, by convention, is from (+) → (−) (i.e., the direction a small [+] test charge would travel if placed in the vicinity of the charge[s] being studied).

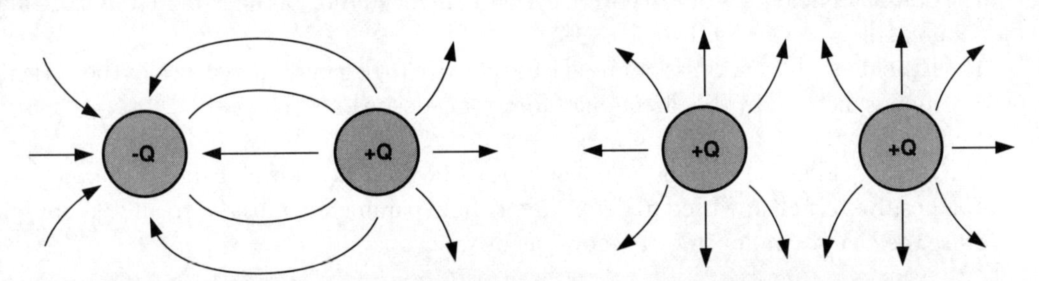

Figure 13-7. E-fields associated with oppositely charged (left) and positively charged (right) objects.

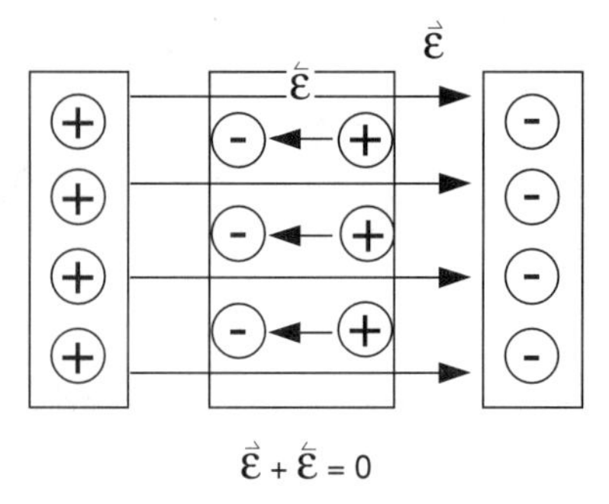

Figure 13-8. Net E-field inside a hollow conductor placed within an external E-field is zero.

To determine the net E-field at a location near two or more independent charges: (1) determine the magnitude of the E-field produced at that location by each charge; and then, (2) sum the individual E-fields using vector addition. To calculate the force a known charge (Q′) would experience if placed in an E-field of known magnitude (E), use the expression:

$$\mathbf{F = Q'E}$$

CONCEPTUAL QUESTIONS

1. What is the magnitude and direction of the E-field at the center of a hollow metal sphere with a radius of 1 m and a surface charge of +4mC?

   Answer: No computation is required. By placing the positive test charge at the center of a circle, perfect symmetry exists; the test charge would not experience a net electric force. Opposing forces balance, and thus, E = 0 within the sphere.

2. What is true about the net E-field inside a hollow metal conductor that is placed within an external E-field?

   Answer: Most likely, the net E-field is zero. In Figure 13-8, note how the external charge rods set up an E-field directed to the right (from positive to negative terminals). Charges within the metal conducting plate are attracted toward the charged rods, creating an E-field directed to the left. Because E-fields are vector quantities, and the E-fields associated with the external charged rods and metal conducting plate oppose one another, the net (resultant) E-field is zero. The E-fields effectively cancel one another.

Remember these important points about E-fields:

1. E-fields exist around charged particles/objects/bodies (although they are "invisible").
2. E-fields have both magnitude and direction.
3. The density of field lines drawn correlates with the magnitude of the E-field (the more lines per area, the stronger the E-field).
4. The E-field, by convention, is oriented from positive to negative charge locales (because the direction is dictated by the direction of force on a positive test charge).
5. Units are N/C.
6. The net E-field inside a hollow conductor placed within an external E-field is zero.
7. The positive test charge used to determine E-field parameters must be small so it does not cause the charges under study to move in any way.

ectrical

14 Electrical Potential

potentia

This chapter is a review of the basic concepts relating to electrical potential energy (PE_q). Understanding these concepts is made easier by familiarity with the gravitational potential energy (PE_g) concepts reviewed in other chapters. Learning by analogy is emphasized.

I. Electrical Potential Energy

Gravitational potential energy (PE_g) is created when a mass is lifted above the ground to the top of a "physical" hill (Figure 14-1**A**). As stated in Chapter 5, the amount of potential energy created depends on the magnitude of the mass, the magnitude of the parameter (g) causing the object to fall toward the ground, and the distance between the top of the hill and the ground surface. A "real" force (F_g) is created as the mass accelerates toward the ground surface. As discussed in Chapter 13, electrical charges also exert "real" force and, as shown subsequently, possess energy (i.e., electrical potential energy, PE_q) .

Electrical forces (F_q) are created when charges are placed near one another (Figure 14-1**B**). A small positive charge placed near a large fixed positive charge experiences a repulsive electrical force, which allows for the creation of a "potential" energy situation. When located near the fixed charge, the mobile charge is said to have maximal **electrical potential energy (PE_q).** As the mobile charge moves away from the fixed charge, the electrostatic force between the charges decreases, as does the induced PE_q. The reduction in PE_q, as the distance between the like charges increases, can be likened to the small charge falling down an "electrical" hill (Figure 14-1**C**).

The **electrical potential (V)** existing at some point A is equal to the amount of work (W) required to bring a positive test charge (+q) from infinite distance away from the charge to point A, divided by the magnitude of the test charge.

$$\mathbf{V = W/q}$$

The electric potential energy (EPE) is equal to the work (W) associated with this process.

$$\mathbf{EPE = W}$$

Thus,

$$\mathbf{V = EPE/q}$$

Figure 14-2 is a diagram of electrical potential (V). Note that the electrical potential at point A is equal to the work to bring the positive test charge from infinity to point A.

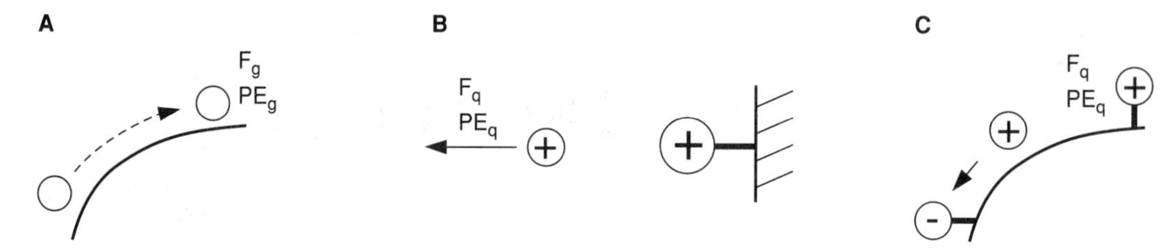

Figure 14-1 Basic concepts in electrical potential.

The potential at a distance r from an isolated point charge Q, or from the center of a uniformly charged sphere, is proportional to the magnitude of the charge (Q) and inversely proportional to the distance r (k, as previously, equals 9×10^9 N·m²/C²). The SI unit of electrical potential is the volt (V) =1 joule/coulomb (J/C). The derivation of the expression for potential is as follows:

$$\textbf{V (potential)} = \frac{W}{{}^{+}q(\infty \rightarrow A)} = \frac{[(F)(s)]}{{}^{+}q} = \frac{[k \cdot Q)({}^{+}q)\ (r)]}{(r^2)\ {}^{+}q} = \frac{kQ}{r}$$

This expression shows the potential associated with bringing a unit positive charge from an infinite distance up to a distance r from a point charge.

A charge's potential is defined in terms of **the amount of work required to move a small, positive test charge from a reference location infinite (∞) distance from the point of interest (with zero absolute potential) to some point at or near the charge of interest. Voltage is a scalar quantity,** having magnitude only.

II. Electrical Potential Difference

To compare the potential energy at two different points requires determining the amount of potential difference between two points, A and B.

The amount of electrical **potential difference** (also referred to as **voltage** or $\mathbf{V_{AB}}$ or **ΔV**) between points A and B is defined as the **work required to move a positive test charge ($^{+}$q) from point B to point A, divided by the magnitude of the test charge** (see Figure 14-2).

The potential difference (i.e., voltage) between points A and B is the magnitude of the potential at point B minus that at point A:

$$\textbf{Potential difference (voltage} = V_{AB} = \Delta V) = V_B - V_A = \frac{W}{{}^{+}q_{(B \rightarrow A)}}$$

$${}_{B}\bullet \text{—}W \rightarrow \bullet_{A}$$

Therefore, the electrical potential energy difference (voltage) between points A to B is the change in electrical potential energy that occurs when a small positive test charge moves from po-

$$\infty \ \oplus \text{——} W \text{———}\blacktriangleright \bullet\ A$$

Figure 14-2. Electrical potential at point A is equal to the work required to bring a positive test charge from infinity to point A, divided by the magnitude of the test charge.

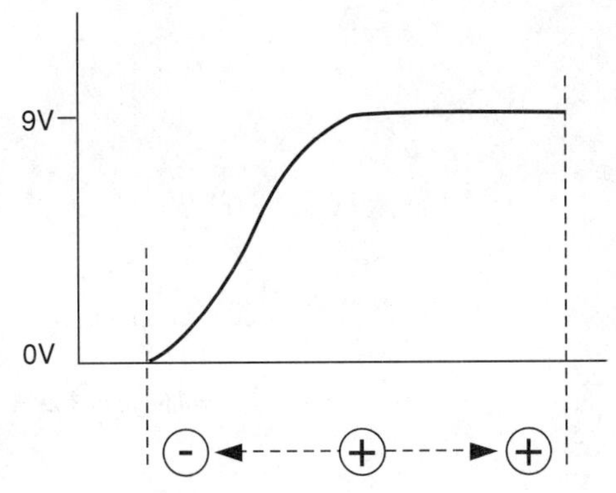

Figure 14-3. Analogy between gravitational and electrical potential energies.

sition B to position A. Because potential difference (voltage) merely refers to a difference in potential, units are also **volts.** A volt is the amount of potential difference between two points if one joule of work is required to move one coulomb of charge from one point to the others.

Remember: Although both potential and potential difference units are expressed in volts, they differ in that potential refers to the amount of potential electrical energy relative to some zero potential reference point, whereas voltage refers to the difference in electrical potential energy between two points of interest of known electrical potential. Whereas electricians use potential difference and voltage interchangeably, a student of physics should use only the term potential difference, be it to describe electrical potential between points A and B, the intracellular and extracellular environments of a human muscle cell, or the microscopic regions between adjacent human neurons.

Potential difference allows for muscle contraction, nerve conduction, and ion transport in tissues. It is what makes a battery a "good" battery. The potential difference between the two terminals of a transistor radio battery is 9 volts. Therefore, the battery's voltage (i.e., difference in electrical potential between the two terminals) is 9 volts. The voltage is the electromotive force (EMF) that drives current (electrical charges) around the closed electrical circuit (see Chapter 16 for more discussion).

Figure 14-3 shows the analogy between gravitational and electrical potential energies. Note that pushing a positively charged ball toward another positively charged ball results in an increase in potential energy, similar to the action of pushing a metal ball up a steep hill. Potential energy is at a minimum as the positively charged ball moves to the left toward the negatively charged ball, as is the potential of a rolling ball at the bottom of a steep hill. In both scenarios, potential energy is a function of relative position (how close to other charged objects or vertical distance from the ground).

The potential difference equation can be modified to allow calculation of the amount of work required to move a charge Q from some point A to B:

$$\mathbf{W = (Q)(\Delta V)}$$

In a uniform electric field (E-field):

$$\mathbf{\Delta V = W/q = [(F)(d)]\,/q = (E)(d)}$$

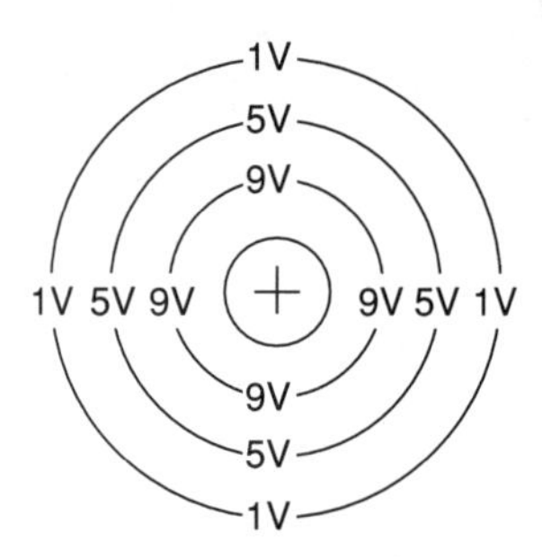

Figure 14-4. Cross section of a positively charged ball surrounded in 3-D by equipotential rings.

in which d is the distance between the two points of interest. Note that potential difference is a **scalar quantity.**

III. Equipotential Lines

The concepts of electrical potential and electrical potential difference (voltage) can be understood by using a pictorial example. A large, positively charged metal ball sits in the plane of this paper. If you place positive test charges around the ball, regions of similar electrical potential would exist around the ball at points equidistant from the ball's center.

If a region of 5 V potential exists at a radius 4 mm away from the ball's center on its left side, the same potential must exist at a radius of 4 mm from its right side, above it, and below it. Thus, **equipotential lines** or **rings** exist around the ball. Every point on a line or ring is of the same electrical potential. Because the metal ball is three dimensional (3-D), 3-D **equipotential surfaces** surround the ball like "rings of onion surrounding the onion core."

A bird's-eye view of a charged ball and the surrounding equipotential rings (i.e., lines) is provided in Figure 14-4. Remember that these rings (lines) are 3-D surfaces that symmetrically surround the charged object.

Because equipotential lines (surfaces) are of the same electrical potential energy, **no work is required to move from one point on a line (surface) to any other point on that same line (surface).** As expected, work is required to move from a line (surface) of low potential to one of higher potential, just as work is required to move from the bottom to the top of a steep hill. Recall that $W = (Q)(\Delta V)$.

The E-field direction is found by seeing what happens to a positive test charge placed in the vicinity of the charges in question. In Figure 14-4, a positive test charge would be moved away from the ball, perpendicular to the equipotential rings. Thus, the E-field direction is perpendicular to equipotential surfaces.

Electromagnetism 15

Qualitative and quantitative understanding of basic electromagnetic (EM) principles and phenomena is important in preparing for a career in basic science, engineering, or medicine. The goal of this chapter is to review key elements of this straightforward and interesting topic that many students have described as "confusing, vague, nebulous, and boring."

I. Magnetic Properties

Some materials have magnetic properties (e.g., magnets) and some do not (i.e., notebook paper). It is the existence of **moving unpaired electrons** that ultimately is responsible for the creation of magnetic properties. Magnetism is a property of some materials composed of certain atoms the electrons of which produce tiny net magnetic fields (B-fields) because of their specific **orbital motion** and **intrinsic magnetic dipole moments.** An unpaired, orbiting electron is like a tiny bar magnet that creates a tiny B-field. Figure 15-1 shows a moving electron creating magnetic field lines like a tiny bar magnet. A standard magnet with its associated magnetic field lines is also shown.

Materials can be classified as having one of three possible magnetic properties:

A. NONMAGNETIC MATERIAL

Electron spin and orbital motion cancel, so this material has no magnetic properties: no B-field is produced and no alignment is seen when the material is placed into a preexisting external B-field.

B. PARAMAGNETIC MATERIAL

Magnetic regions or "domains" exist in the material. The net B-field of the material is about zero, although alignment of "**magnetic domains**" is seen when the material is placed into an external B-field. A weak, temporary B-field exists when placed into a preexisting external B-field. Figure 15-2**A** shows an object that con-sists of many regions or domains, each with its own magnetic dipole moment. In Figure 15-2**B,** the B-fields of each domain tend, for the most part, to align with the direction of the external B-field while it is present. In Figure 15-2**C,** the near total domain alignment in Figure 15-2**B** is lost soon after the object is removed from the external field.

Thermal motion of unpaired electrons is responsible for destroying alignment of B-field domains. Most of the **transition metals** and their compounds in oxidation states involving incomplete inner electron subshells are paramagnetic.

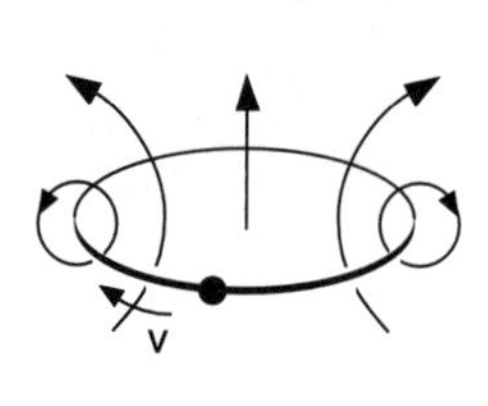

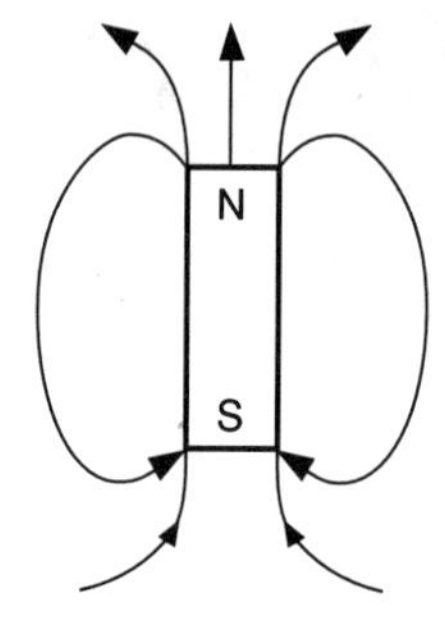

Figure 15-1. Magnetic properties of materials depend on moving unpaired electrons. At left, an electron moving in a circular orbit and the resulting magnetic field lines. At right, the magnetic field lines associated with a magnet.

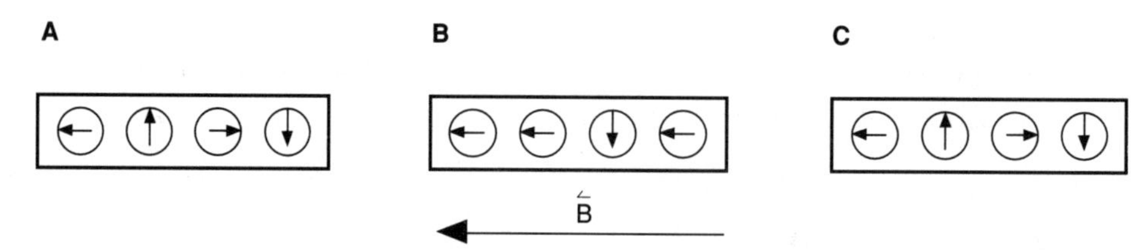

Figure 15-2. An object consisting of domains, each with its own magnetic dipole moment. Note the effect of an external B-field on domain dipole alignment.

C. FERROMAGNETIC MATERIAL

Materials consisting of magnetic "domains" retain their alignment after being removed from an external B-field. These materials subsequently have a net B-field associated with them; thus, they are, or can be, **permanently** "magnetized." Figure 15-3 illustrates how domains within ferromagnetic materials align with an external B-field and retain their alignment even after the external field is removed. Ferromagnetism can be considered an extreme form of paramagnetism. A critical interatomic distance (not too close, not to far) is required. These materials tend to consist of elements with incompletely filled *d* or *f* electron subshells. Materials composed of iron (Fe), cobalt (Co), nickel (Ni), and gadolinium (Gd) are ferromagnetic at room temperature.

II. Magnetic Fields

A positive test charge exists in space. If you detect a force acting on this test charge when it is at rest, an electrostatic field is present. If the test charge is moving, it experiences a magnetic force and a magnetic field exists.

Magnetic fields (B-fields) are **real** fields (like gravitational and E-fields) that can be "emitted" from an object and create "forces" that cause objects nearby to move in an altered path. B-

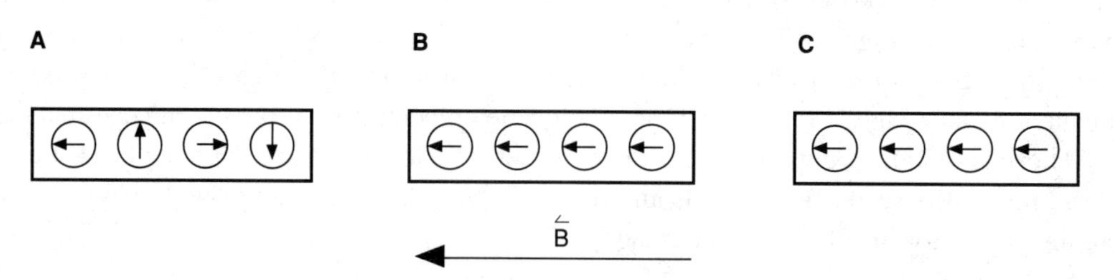

Figure 15-3. Alignment of domains within ferromagnetic materials due to an external B-field.

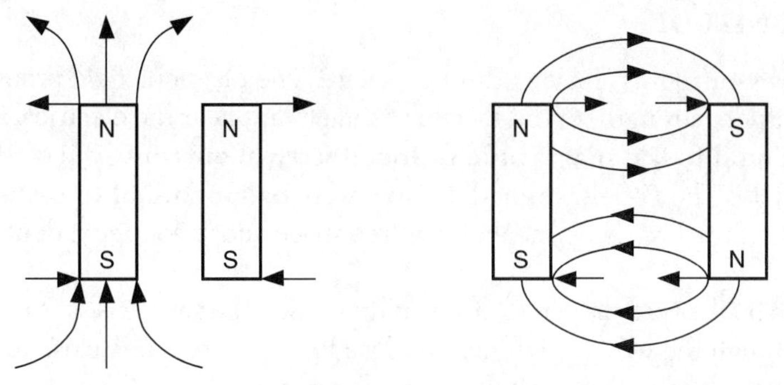

Figure 15-4. Magnetic fields as an "intrinsic" property. Here, magnets "emit" B-fields. At left, two magnets with similar poles close together emit fields that oppose one another. At right, two magnets with poles attract one another.

fields are either "intrinsic" properties of a material or object, or they can be induced by moving electric charges.

A. INTRINSIC PROPERTY

Permanent magnets are an example. Magnets, such as those with which children play, have a net magnetic dipole moment associated with them. Electron orbital motion and intrinsic magnetic moment are ultimately responsible for the object "emitting" B-fields (Figure 15-4). Like gravitational (G) and electric (E) fields, B-fields have both magnitude and direction. **B-fields are vector quantities.** The magnitude can vary (the more lines per unit area, the greater the magnitude); **the direction is always from north to south poles.** As with charges, like poles repel, opposite poles attract.

B. INDUCED PROPERTY

Electrical wires are an example. Think of a segment of electrical wire as a substance (a metal) through which electrons travel. The electrons traveling along the length of wire are responsible for the creation (induction) of B-fields around the wire (Figure 15-5). B-fields emanate from the wire, just as sound waves emanate from a loud speaker, and "fade away" (decrease in magnitude) as one moves farther and farther from the source (the electrical wire). "Induced" B-fields are produced as a result of charge moving through a metal wire.

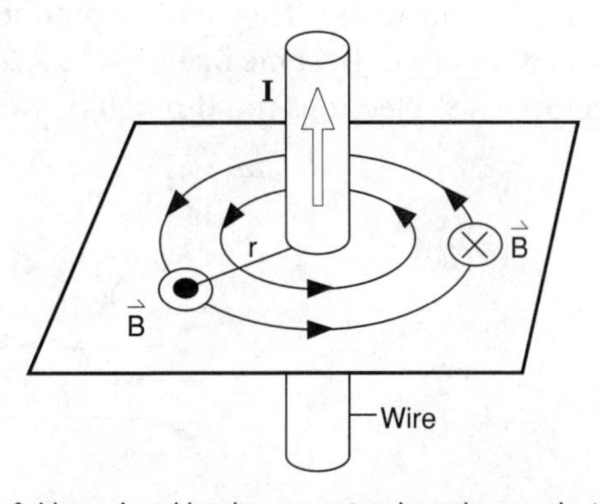

Figure 15-5. "Induced" magnetic field, produced by charge moving through a metal wire. "Circle" indicates the magnetic field comes out of the plane, while "X" indicates magnetic field goes into the plane.

1. MAGNITUDE

Figure 15-5 shows current (I) moving through a wire. The magnetic field produced encircles the wire. The magnitude (strength) of the B-field at some point near the electrical wire should **be directly proportional to the magnitude of the electrical current** (I) [i.e., the flow of charge] producing the field. The B-field strength **is inversely proportional to distance (r) from the wire.** The constant (m_o) is the permeability of free space and is not dependent on the material or wire.

Like sound intensity, B-field magnitude is high when the source is "at a maximum" (i.e., current running through the wire is high) and is close to the source (i.e., close to the wire). Magnitude can be expressed as:

$$\mathbf{B} = \frac{\mathbf{u_o\, I}}{\mathbf{2\pi r}}$$

The SI unit of B-field magnitude is the **Tesla (T)** = Weber/m^2 = 10^4 gauss. The B-field produced by the common toy magnet is about 0.005 T; that produced by the planet Earth is about 5×10^{-5} T.

Magnets produce B-fields that can do many things. Strong B-fields erase credit cards or ATM cards. The magnets used in magnetic resonance imaging of the body produce a B-field of about 1 to 2 T.

The concept that current moving through an electrical wire sets up a surrounding B-field is important to understand because it holds true for wires of many shapes: linear (as just mentioned), loop-shaped (Figure 15-6, at left), and solenoid/coil (Figure 15-6, at right). **The magnitude of the B-field created around the wire is directly proportional to the amount of current and inversely proportional to the distance from the wire to the point of interest.** Therefore, the B-field experienced as a result of current travelling through a wire is maximum when I is maximum and distance from the wire is minimized.

2. DIRECTION

The location and direction of the B-field produced around an electrical wire can be determined by using the **Right Hand Rule Part I,** which says: **If you wrap the fingers of your right hand around the wire and point your thumb in the direction the electrical current (I) travels, your nonthumb fingers (on your right hand) curl in the direction of the induced B-field.** Review Figures 15-5 and 15-6 and make sure you agree with the direction of the B-fields shown by using the Right Hand Rule.

B-fields are real and "alive" in our world. They can be produced by some materials at all times (e.g., magnets) or produced (induced) by some materials (e.g., electrical wires, power lines, appliance cords, overhead lamps) when electrons run through them.

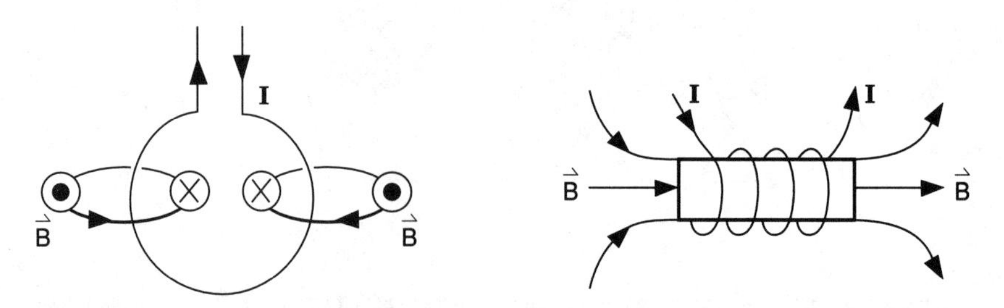

Figure 15-6. B-fields produced by current passing through a loop (left) and solenoid/coil (right).

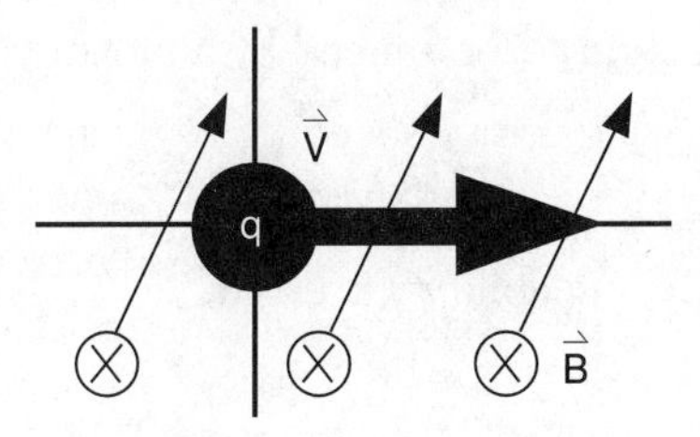

Figure 15-7. Magnetic force (F_{mag}). A charge subjected to a magnetic force moves with speed v, is in the vicinity of a B-field, and moves such that the velocity vector is perpendicular to the B vector.

III. Magnetic Forces

A gravitational field is responsible for the production of a gravitational force, which, in turn, can cause objects to accelerate toward the ground. An electric field is responsible for the production of an electric force, which, in turn, can cause charged objects to move in some way. A magnetic field has the potential to produce a **magnetic force** (F_{mag}), which, in turn, can cause moving charged objects to deviate or deflect from their original course (Figure 15-7).

Magnetic forces are real forces, as they can exert an action force on moving charged particles or objects. A **magnetic force** is created and can act on an object **if and only if:**

1. The object of interest is electrically charged (with a charge **q**)
2. The object of interest is moving (with a speed **v**)
3. The object is in the vicinity of a magnetic field (of magnitude **B**)
4. The object is moving in such a way that:
   a. **its v vector is perpendicular (90°) to the B vector** *or*
   b. **a component of its v vector is perpendicular (90°) to the B vector**

A. MAGNITUDE

The **magnitude** of the magnetic force (F_{mag}) can be determined using the expression:

$$\mathbf{Fmag = q\ v\ B\ \sin \phi}$$

in which ø is the angle (0 to 180°) between the v and B vectors (Figure 15-8).

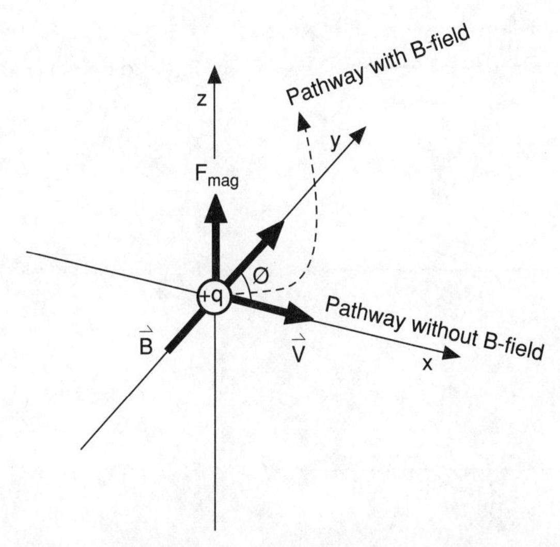

Figure 15-8. Relationships between (F_{mag}), B, v, and ϕ. Deflection of a positively charged particle ($+q$) from its original path (along the x-axis) toward the vertical (x–z plane) is caused by F_{mag}. Note: $+q$ at origin of the figure.

The amount of force (deflection) experienced by a moving charge is proportional to:

1. How much charge is present
2. The speed at which the charge is moving
3. The amount of B-field ("pushing ability") present
4. The "push angle" (ϕ)

Thus, highly charged particles or objects that are moving quickly in a region of a strong B-field are set up for experiencing a strong magnetic (push) force. If all else is equal, F_{mag} is maximal when ϕ=90° (90° is the best "push angle"). As shown in examples 2 and 3 of Figure 15-9, F_{mag} causes the moving charge to deviate from its linear original path along the *x* axis; the charge moves along the *x* axis but soon deflects vertically.

B. DIRECTION

The **direction** of a magnetic force is determined by using the technique called the **Right Hand Rule Part II,** which says: **If you point your right thumb in the direction the charge travels (the direction of the v vector), and your remaining (straightened) right fingers in the direction of the B-field, your palm will face the direction of F_{mag} if the moving object is positively (+) charged. If the object of interest is negatively (−) charged, F_{mag} is simply the exact opposite direction or, simply, the direction the back of your hand faces.**

Be aware of symbols used to note relative direction: velocity and/or B-field vectors oriented into the plane of the paper are symbolized with a cross mark (the view of an arrow tail), and those oriented out of the plane of the paper (coming out directly at you) are symbolized with a bullseye mark (the view of an arrow tip coming toward your face).

In the examples in Figure 15-9, determine if an F_{mag} exists in each case.

If a charge moving at constant speed enters a B-field with its velocity perpendicular to the B-field, the charge will be deflected in a circular orbit. The radius of this orbit can be determined using the concept that magnetic force is converted to centripetal force:

$$\mathbf{F_{mag} = F_{centripetal}}, \qquad \mathbf{qvB = (mv^2)/r}, \qquad \mathbf{r = (mv)/qB}$$

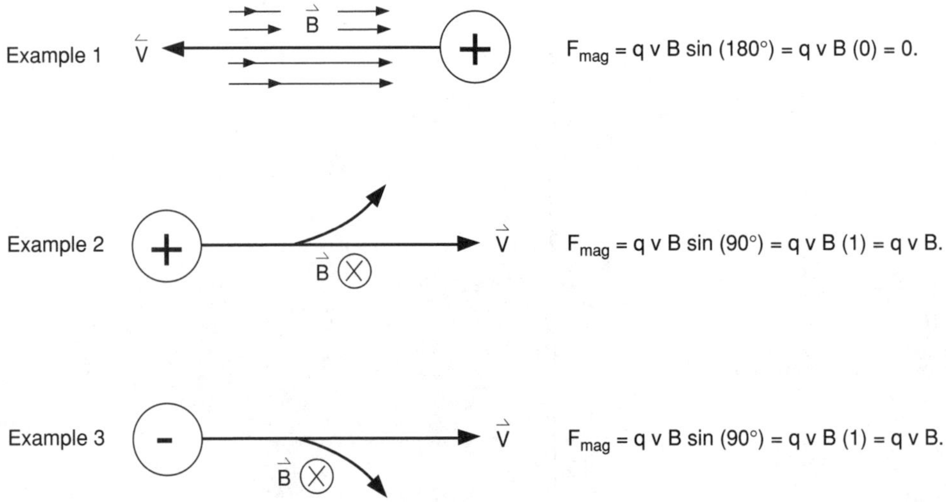

Example 1: $F_{mag} = q\,v\,B \sin(180°) = q\,v\,B\,(0) = 0.$

Example 2: $F_{mag} = q\,v\,B \sin(90°) = q\,v\,B\,(1) = q\,v\,B.$

Example 3: $F_{mag} = q\,v\,B \sin(90°) = q\,v\,B\,(1) = q\,v\,B.$

Figure 15-9. Three examples showing the calculation of F_{mag}. A charge with a velocity and a B-field are shown in each case. Note the deflection of the charge in Examples 2 and 3 caused by F_{mag}.

Remember, no magnetic force is created ($\mathbf{F_{mag} = 0}$) for the following:

1. Uncharged particles or objects
2. Charges at rest
3. Charges moving in a region with no B-field present
4. Charged particles moving parallel to the B-field

A magnetic force is produced **only** if the conditions are right—**a charged particle or object is moving through a B-field such that the v and B vectors have some perpendicular orientation to each other.** Note also that because the magnetic force vector is always perpendicular to the velocity/displacement vector ($\theta = 90°$), **no work is done by magnetic force on the moving charge (i.e., W = F s cos 90° = 0).**

C. MAGNETIC FORCES COMPARED WITH GRAVITATIONAL AND ELECTRICAL FORCES

Magnetic forces are fundamentally different from gravitational and electric forces. Although magnetic force is exerted only on objects that are moving, gravitational and electric forces are exerted on objects that are moving and on those at rest. Whereas magnetic force exists only if the direction (or a component) of the charge's motion is perpendicular to the B-field, gravitational and electric forces are independent of the direction in which the object moves.

More complicated scenarios involving magnetic force include:

1. The magnetohydrodynamic generator (E- and B-field balance)
2. The mass spectrometer (circular motion in a magnetic field)
3. The electric motor (wire loop magnetic torque principle)

No matter the scenario, moving charged particles or objects may be deflected from their original (usually linear) path on entering a region of uniform magnetic field. The direction of deflection is determined by the Right Hand Rule Part II [the right palm faces the direction of F_{mag} for positive (+) charges, the back of the hand faces the direction of F_{mag} for negative (−) charges].

IV. Magnetic Flux

A solitary loop of metal wire (Figure 15-10) initially has no electrical current or magnetic field running through it or any involvement with any outside process. A bar magnet with known B-field strength (B) is brought from the right toward the loop (North pole of the bar magnet facing the

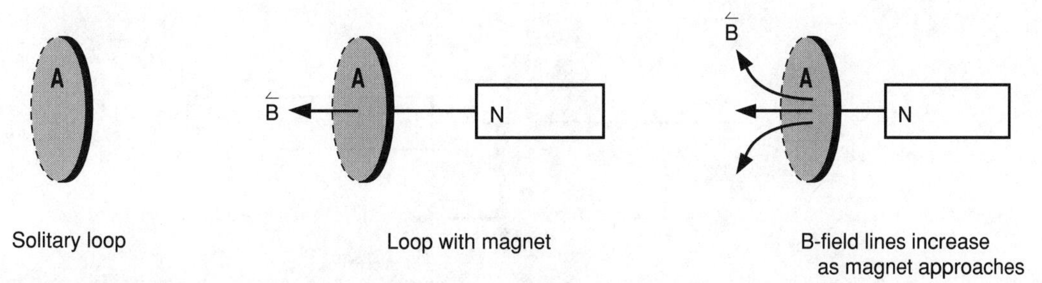

Figure 15-10. Magnetic flux. Note the B-field lines as a magnet approaches a metal loop. The amount of magnetic field within the loop at a specific time is the magnetic flux.

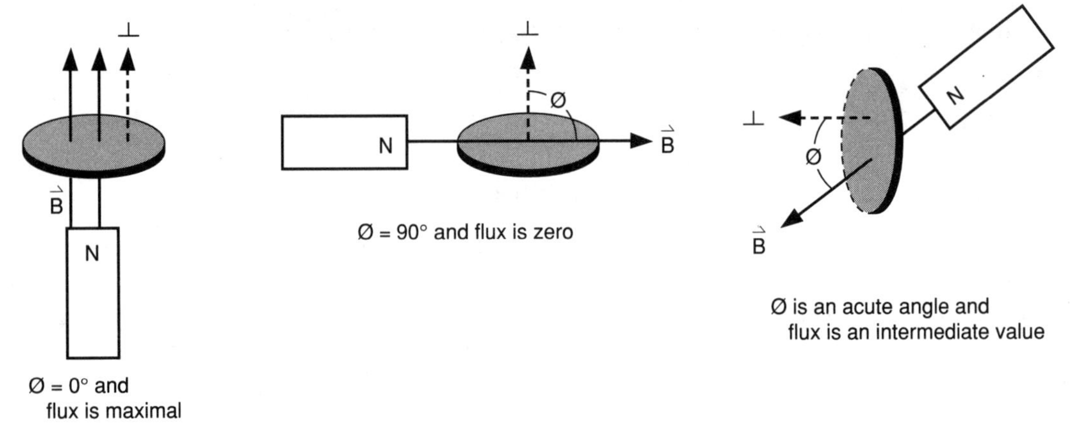

Figure 15-11. Magnetic flux with differing angles between B and a line perpendicular to the plane of the loop (differing ϕ values).

loop). B-fields from the magnet pass through the loop as the magnet approaches the loop. More and more B-field lines pass through the loop as the magnet is brought closer to the loop.

The amount of magnetic field within the loop at a specific time is called the **magnetic flux (F).** The magnetic flux through the loop is proportional to the amount of area inside the loop, the strength of the B-field within the loop, and the orientation of the loop in the field. A quantitative expression is:

$$\textbf{Magnetic flux } (\Phi) = \mathbf{A\ B\ cos}\ \phi$$

in which A is the area inside the loop through which the external B-field passes, B is the magnitude of the magnetic field passing through the loop, and ϕ is the angle between B and a line perpendicular to the plane of the loop. Note that Φ is maximum when ϕ is zero degrees (i.e., the B-field lines fully transverse the area of the loop). Flux is zero when $\phi = 90°$ (because $\cos 90° = 0$). These concepts are illustrated in Figure 15-11.

For purposes of understanding basic concepts, think of magnetic flux as a "bad" process, one that needs to be corrected (i.e., reversed). The external B-field ($B_{external}$) introduced disturbs the state of "homeostasis" in which the loop had been (it was used to having **no** magnetic fields present). A metal loop can return to its preflux state by counteracting the flux itself. The external B-field lines introduced by the magnet are negated or "neutralized" by the loop. The loop creates, or induces, its own B-field ($B_{induced}$) directed exactly **opposite to the change in the B-field** introduced by the magnet ($B_{external}$) (Figure 15-12).

Question: If you were a metal loop, how could you produce a B-field?
Answer: Produce your own. A B-field can be induced in a metal, conducting hoop by allowing current to run through it. Thus, a B-field can be induced by producing an **"induced" volt-**

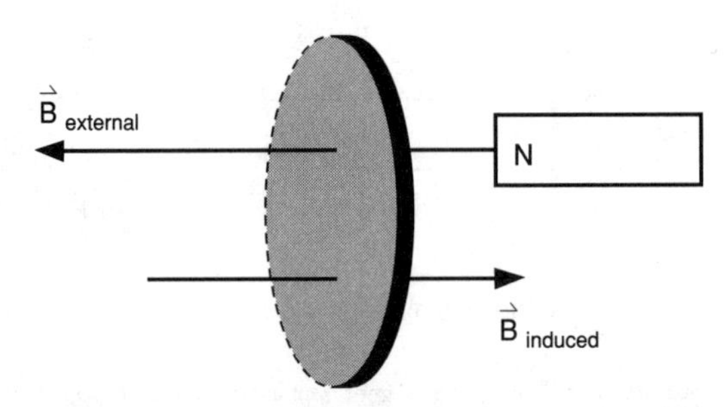

Figure 15-12. Induction of a B-field by a loop ($B_{induced}$).

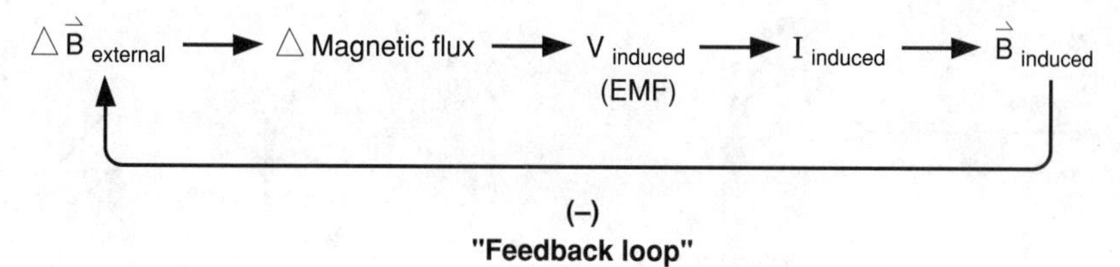

Figure 15-13. An EMF induced within a conducting loop causes a current to flow within the loop. Lenz's law says that an induced EMF produces a current that tends to oppose the change of flux that caused it. Note the "feedback" loop.

age ($EMF_{induced}$), which in turn creates an **"induced" electrical current ($I_{induced}$);** it is $I_{induced}$ that is responsible for the creation of $B_{induced}$, which counters the B-field lines emanating from the external magnetic field **($B_{external}$).**

The process by which voltage is induced by varying magnetic field inside a loop of wire is **electromagnetic induction;** it was discovered by Faraday in 1831. **Faraday's law says that the induced voltage (EMF) through a loop of wire equals the change of magnetic flux ($\Delta\Phi$) through the loop divided by the time (Δt) needed for that change in flux, or:**

$$\epsilon = -\Delta\Phi/\Delta t$$

SI units are as follows: 1 Wb/s = 1 V. The negative sign indicates that the induced voltage *opposes* the changing flux that caused that voltage.

Lenz's law states that the polarity of an induced voltage is such that it produces a current the magnetic field of which *opposes* the change in flux that caused the induced voltage. Electromagnetic induction can be viewed as a series of individual physical processes that "work together to solve a problem" (note the "negative feedback" loop—a term borrowed from the biological sciences) [Figure 15-13].

A **change** in the flux "induces" this chain of events. A static loop in a constant B-field will **not** have an induced voltage. Only during the time the B-field, and therefore the flux, changes are an induced voltage, current, and B-field established.

You can use the Right Hand Rule Part I to determine the polarity (i.e., direction) of the induced current on the metal loop. Knowing that $B_{induced}$ should point in the direction to oppose the change in flux created as a result of some experimental procedure (e.g., pushing a bar magnet toward the loop, pulling the bar back through the loop, suddenly increasing or decreasing the magnitude of the $B_{external}$), curl your right fingers around the loop to curl in the direction of the required $B_{induced}$. The direction your right thumb points while grasping the loop (with fingers curling the needed $B_{induced}$) gives the direction of $I_{induced}$ on the wire loop.

Figure 15-14 offers the specifics for the example of magnetic flux in Figures 15-10 and 15-11. Note that $B_{induced}$ points toward the right to counter the $B_{external}$, which is directed to the left. Using Right Hand Rule Part I, have your right fingers curl and fingertips point toward the right. If you grasp the segment of the loop projecting out of the plane of the paper (sticking out at you), your right thumb points downward. Thus, $I_{induced}$ travels counterclockwise around the loop if you are an observer standing just to the right of the loop (see Figure 15-14).

V. Electromagnetic Spectrum

An **electromagnetic (EM) wave** is an **electrical and magnetic disturbance** that moves through space (or a medium) at a definite speed. All waves move through space (a vacuum) at the

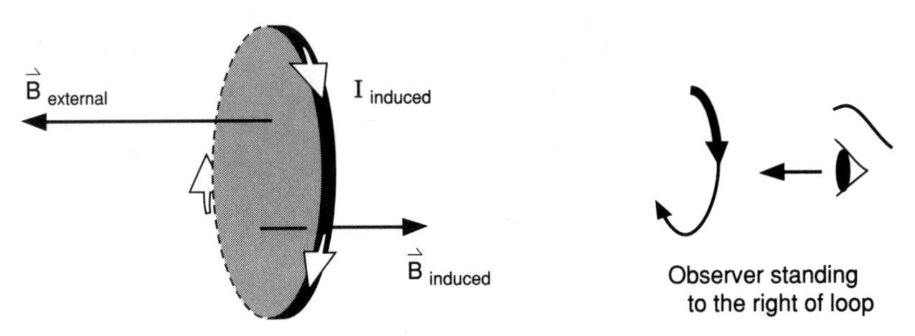

Figure 15-14. The Right Hand Rule, Part I is used to find the direction of $I_{induced}$.

speed of light (c, c = 300 million meters/second = 3×10^8 m/sec). Electromagnetic waves vary in energy content; wave energy depends on its **frequency (f)** of vibration (f is the number of complete vibrations per second of the field at a point along the path of the passing wave). Electromagnetic waves with "high" energy content (e.g., x-rays) vibrate more times per second than waves with "low" energy content (e.g., radio waves). The actual length of one wave (i.e., one complete vibration pattern) is the wave's **wavelength (λ). The following relationship holds true for all EM waves travelling in a vacuum.**

Speed of EM wave travel (c) = Frequency of wave (f) × wavelength of wave (λ)

$$c = f\lambda$$

Electromagnetic waves are usually classified by frequency of vibration in a scheme called the **electromagnetic (EM) spectrum.** Examples of EM waves in increasing frequency/energy content and decreasing wavelength are: (1) AM and FM radio waves; (2) microwaves; (3) infrared light; (4) visible light; (5) ultraviolet light; (6) x-rays; and (7) gamma rays.

The EM wave is ultimately produced by an oscillating electric charge. Consider a radio station antenna (Figure 15-15). Electric fields (E-fields) and magnetic fields (B-fields) are **generated** then **radiate** from the antenna because of the continued movement of electrical charge up and down the length of the antenna. The moving charges in the antenna are responsible for producing both E-fields and B-fields that originate right next to the antenna. Fields at a specific location near the antenna have a specific magnitude and direction at some specific time. The mag-

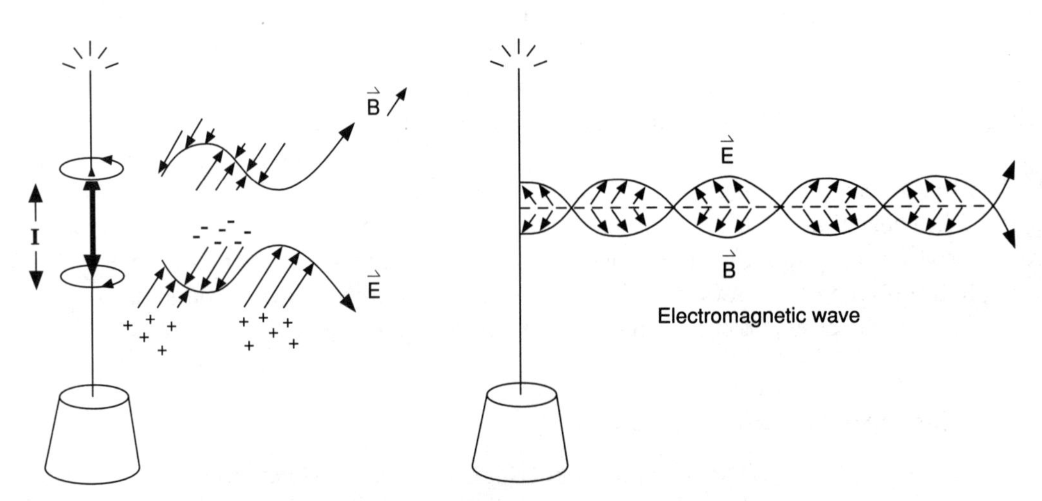

Figure 15-15. Wave forms seen over time as charge moves up and down antenna.

nitude (strength) of the fields is determined by the relative magnitude and location of the charge in the antenna. The polarity (i.e., direction) of the fields produced around the antenna varies in orientation (up [+] or down [−]) with the direction the charge moves within the antenna itself. The magnitude and polarity of the E- and B-fields existing at any one point near the antenna varies with respect to time. This variation (E- and B-field magnitude and polarity) over time is responsible for the production of the "wave" form. The waves move at a speed "v" away from the antenna. The combined electrical and magnetic field moving at speed c is an **electromagnetic wave.**

ircuits

DC Circuits

16

dc circu

This chapter brings electronic principles together. The DC circuit, a collection of various different separate entities, is defined and studied. Important topics include: batteries (the source of EMF that drives the circuit and creates voltage), current (the flow of electrical charge around the circuit), resistors (devices that resist the flow of electrical charge and consume voltage), capacitors (devices that store electrical charge, for possible later use), dielectrics (materials that allow capacitors to hold additional electrical charge), electric power (the rate at which electrical energy is converted to other forms of energy), and DC circuit physics.

I. Terms and Concepts

A. DC AND AC CIRCUITS

Direct current (DC) electric circuits are closed electrical systems in which the current (I, the flow of charge) and the voltage (V, the potential difference that allows charge to flow) are **constant** over time. A DC circuit is inside a transistor radio. A transistor radio battery produces a constant current and voltage. Another type of circuit—the **alternating current (AC)** circuit (see Chapter 18)—differs from the DC circuit in that the current and voltage in AC circuits **vary** over time as a sinusoidal wave function (Figure 16-1). The electrical circuit in most homes is an example of an AC circuit.

B. BATTERIES

Batteries separate positive and negative charge, thus creating a **voltage** (potential difference) across them. Batteries are the source of **electromotive force (EMF)** that drives the circuit. Batteries convert chemical energy into useable electrical energy. "Good" batteries are able to separate charge, whereas "bad" batteries are unable to do so and should not be used because they can no longer drive a DC circuit. The common transistor battery has a voltage of 9 volts, which means the potential difference between the two battery terminals is 9 volts. The symbol for the battery is shown in Figure 16-2.

C. CURRENT

Current (I) refers to the rate at which charge flows past a point in the circuit:

$$I = \Delta Q/\Delta t$$

The unit of current is the ampere. **1 amp of current = 1 Coulomb of charge flowing past some point on the wire/1 second.** Current is really the flow of electrons through a con-

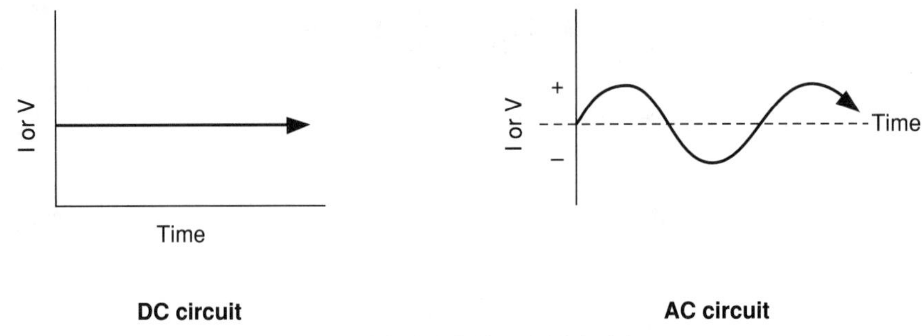

Figure 16-1. The DC (left) and AC (right) circuits.

ducting material such as a wire. Because of certain conventions, however, the direction of current is **not** considered the direction electrons flow, but rather, the direction that positive (+) charge flows. Thus, current is "**conceptualized**" **as being the direction positive (+) charge flows** (i.e., current always flows from a point of higher potential to a point of lower potential). Use this convention whenever dealing with both conceptual and mathematic problems related to current.

D. RESISTANCE

Resistors are materials or deliberately designed devices that oppose the flow of electric current. They resist the flow of current by consuming voltage and releasing it in the form of **heat.** Resistors provide an example of how one form of energy (electrical potential energy) can be converted to another form (thermal energy). The unit of resistance is the ohm (Ω), named after the 19th century German electrical physicist Georg Ohm. The symbol for a resistor is .

One **ohm is the electrical resistance of an object that will allow a 1-amp electrical current to flow when a 1-volt potential difference is placed across the object.** Resistance of a metal wire, or any body for that matter, depends on its length (L; the longer it is, the more resistance it has), cross-sectional area (A; the greater the area, the less resistance), and resistivity (r; an intrinsic measure of how well a material resists the flow of charge; the greater the value, the greater the resistance):

$$\mathbf{R = \rho\ (L/A)}$$

E. KEY RELATIONSHIPS

Ohm's law says that the three quantities just described (V, I, and R) are related as follows:

$$\mathbf{V = I \cdot R}$$

This equation is used to calculate the voltage in a DC circuit if you know its total current and total resistance. Likewise, it is possible to determine the total current flowing through a DC circuit when the voltage of a battery (source of EMF) and total circuit resistance are known. In most test problems, the total circuit resistance is usually related to resistors. The wire itself might

Figure 16-2. The symbol for a battery.

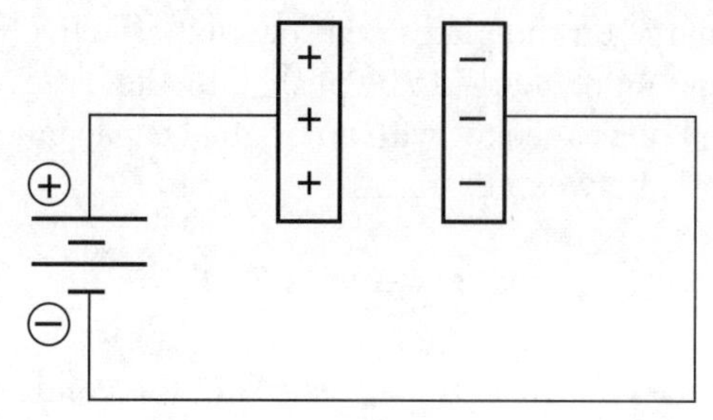

Figure 16-3. A capacitor connected to a battery. The capacitor stores charge on its conducting surfaces.

have a small amount of internal resistance, but usually the amount is so insignificant that it is not considered.

The electrical power used by a device can be determined by using the equation:

$$\textbf{Power} = \mathbf{IV} = \mathbf{I^2R} = \mathbf{V^2/R}$$

Remember that one unit of electrical power, 1 watt (1 J/sec) = 1 volt-amp = 1 amp^2-ohm = 1 volt^2/ohm.

F. CAPACITANCE

Capacitors are the fourth and final component of the DC circuit (the conducting metal wire, battery, and resistor are the first three components). **A capacitor is a pair of conductors: one conductor with a negative charge and the other with an equal positive charge. Capacitors store electric charge on their conducting surface.** The capacitor consists of two electrical conducting plates separated by a vacuum, air, or other nonconducting material (Figure 16-3). A capacitor does not allow electric current to pass through it.

The electronic camera flash device is an example of a capacitor at work; it stores a great deal of charge (the **charging** process often generates a high-pitched "whining" sound) and then releases it suddenly when you push the camera button. The **discharging** process provides a great deal of charge quickly that can be used by the flash light bulb to create a bright flash. The symbol for the capacitor is: ⊣⊢.

Capacitance (C) is the parameter that indicates how "good" a capacitor really is; that is, how much charge it can hold. The **better the capacitor, the higher its capacitance and the more charge it can hold on each of its two plates. Capacitance is defined as the charge on either conductor (plate) divided by the potential difference between the conductors:**

$$\mathbf{C = Q/V}$$

When an insulating material is placed between the two conductors of a parallel plate capacitor, the capacitance increases. The insulating material is said to have a **dielectric constant K,** which is a measure of how much the insulating material increases the capacitance compared to when a vacuum exists between the capacitor plates:

$$\mathbf{K = C_{insulator}/C_{vac}}$$

In short, dielectrics allow the plates to hold more charge; thus, the capacitance (i.e., the charge-storing ability) of the capacitor is increased. Dielectrics are made of polar compounds that tend to interact with the charge on the plates in a way that effectively "neutralizes" some of the charge on them, which allows more charge to be pushed by the battery onto the plates. Thus, the net or total capacitance (C_T) of a capacitor with an original capacitance (C_o) is increased by having a dielectric (with a $K > 1$) between its plates:

$$\mathbf{C_T = K \cdot C_o}$$

Capacitance also depends on the surface area (A) of the conducting surfaces, the distance separating the two plate surfaces (d), the dielectric constant (K) of the material between the plates, and a constant (ϵ_o):

$$\mathbf{C = K \cdot \epsilon_o \cdot (A/d)}$$

Note that capacitance is maximum for large plates that are close together but separated by a material with a large dielectric constant K, which allows for the storage of maximal electrical charge.

The unit of capacitance is the farad, named in honor of the English electrician Michael Faraday. 1 farad = 1 coulomb/volt. A 1-farad capacitor attached to a 1-volt battery would have a charge of +1 C on one plate and −1C on the other. Typical capacitors have a capacitance on the order of the microfarad ($\mu F = 10^{-6}$ F).

The change in electrical potential energy that occurs when a capacitor (C) is charged with a charge (Q) from a battery with potential difference (V) is equal to:

$$\mathbf{1/2\ QV = 1/2\ CV^2 = 1/2\ Q^2/C}$$

II. Physics of the DC Circuit

The DC circuit is a collection of various electronic elements: conducting wire, battery, resistor(s), and/or capacitor(s). The backbone of the circuit is the metal conducting material (usually metal wire) connecting the various circuit components. The power horse of the circuit is the battery, the source of voltage (EMF). One or more resistors consume voltage so the circuit will function (the voltage produced by the battery **must** be consumed **fully** for a potential difference to exist between the two ends of the circuit). If the battery produces 9 volts, 9 volts **must** be consumed by the resistor(s) present.

Current, on the other hand, remains constant in a DC circuit, except where the circuit divides into two or more separate junctions. **The sum of the currents within each junction must add to give the total current that was present just before and after the junction.** Figure 16-4 illustrates a simple (series, i.e., no branching occurs) DC circuit composed of a 9-volt battery, a metal connecting wire, and a light bulb with a resistance (R) of 9 ohms.

Note that a potential energy gain occurs across the battery terminals (0 volts exist at the negative terminal, 9 volts exist at the positive terminal). Recall that the direction positive charge follows is the direction of the current. Using Ohm's law, the current (I) produced in this circuit is:

$$\mathbf{I = V/R = 9\ volts/9\ ohms = 1\ amp}$$

Voltage drop (V_D) across the bulb (a source of resistance, voltage is "consumed" and heat is produced) **must** be 9 volts because **all** 9 volts produced by the battery must be "consumed" by the

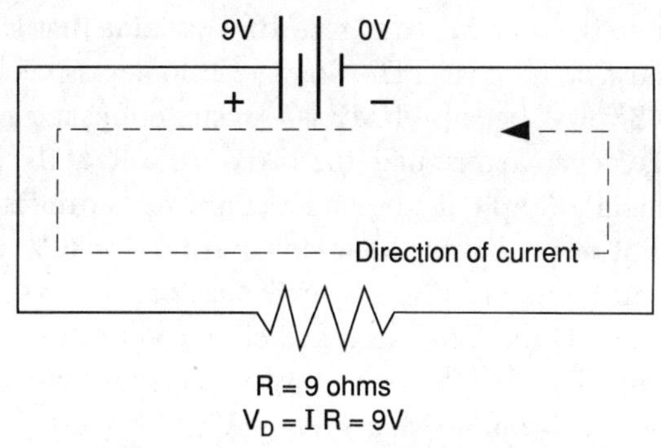

Figure 16-4. A simple series DC circuit.

time you get to the end of the circuit (i.e., the negative terminal) where V must equal zero. The current, 1 amp, is the same at every point in this series circuit.

Figure 16-5 shows a parallel type of connection in which the current reaches a **junction** (i.e., branching) point. The two resistors are said to be **parallel** (i.e., the charge travelling within the circuit must branch and travel to each separate circuit component). The current created by the battery breaks up at the junction in an **inverse ratio to the resistances of the resistors.** Thus, if the current sees two resistors in parallel, R_1 has a resistance of 2 Ω and R_2 has a resistance of 4 Ω (i.e., R_2 has twice as much resistance as R_1), the current splits in a way such that twice as much goes to the resistor of less resistance (R_1) than to the one of greater resistance (R_2).

A circuit may include several resistors. Resistors connected end to end (Figure 16-6, at left) are said to be connected "in **series,**" whereas those connected by a junction (Figure 16-6, at right) are said to be connected "in **parallel.**"

To determine the total resistance (R_T) of resistors in **series,** sum the individual values: $R_T = R_1 + R_2 + R_3 + \cdots$. To determine the total resistance of resistors in **parallel,** solve for R_T using the expression: $1/R_T = 1/R_1 + 1/R_2 + 1/R_3 + \cdots$. You must determine R_T if you want to know the V_T or I_T of a circuit, because Ohm's law says: $V_T = I_T \cdot R_T$.

The voltage drop across any resistor in a solely series circuit (V_{Dx}) is equal to the I travelling through that resistor (I_T) times the resistance of that resistor (R_x). The V_D across resistors in parallel with each other (within a single junction) **must** be equal, and can be determined using Ohm's law: V_D at a resistor in parallel connection is equal to the current (I) travelling through that resistor times the resistance (R) of that resistor.

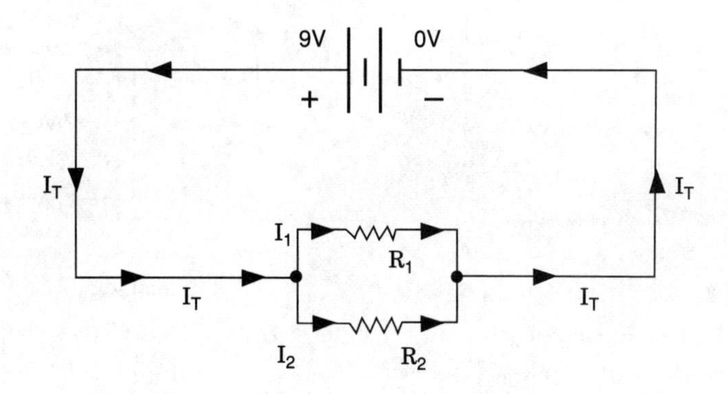

Summary:

$V = I_T \cdot R_T$; $I_T = I_1 + I_2$; $1/R_T = 1/R_1 + 1/R_2$; EMF = 9V = $V_{DR1} = V_{DR2}$.

Figure 16-5. A circuit with parallel resistors.

In the series circuit on the left side of Figure 16-6, assume that EMF = 9 volts, I = 1 amp, and $R_1 = R_2 = R_3$ (R must be 3 ohms). The voltage drop across each resistor must be 3 volts ($V_{DR1} = V_{DR2} = V_{DR3} = 3$ volts), because EMF is "consumed" as it travels along the circuit from the positive terminal to the negative terminal (the EMF is 9 volts at the positive terminal and zero volts at the negative terminal). Simple mathematics confirms our findings: $V_{DR1} + V_{DR2} + V_{DR3} = 9$ volts. **Remember that for resistors in series connections: $V_{DR1} = I \cdot R_1$.**

The parallel circuit in Figure 16-6 gives three resistors (R_1, R_2, and R_3) in parallel within one junction. Knowing that: (1) the EMF in a DC circuit must be "consumed" by the resistors contained within the circuit ($R_{1\text{-}3}$ are the only resistors present here), and (2) the V_D across resistors in parallel are the same, it follows that $V_{DR1} = V_{DR2} = V_{DR3} = 9$ volts (each voltage drop must be the same, and the junction as a whole must consume the entire 9 volts, by rule). **Note:** if 10 resistors are placed in parallel here, each would have a 9-volt voltage drop across it. Even if 50 resistors are placed in parallel here, each would have a 9-volt drop across it.

The equal voltage drop across resistors of different resistance in parallel can be appreciated by looking at Ohm's law: $V_{Dx} = I_x \cdot R_x$. V_D for resistors in parallel are equal because of a "compensation mechanism," i.e., I and R values have an inverse relationship in parallel connections. Think of an example: a resistor with 1 unit of resistance gets twice as much current running through it than a resistor of 2 units resistance; a resistor of 2 units resistance gets one-half the current running through it than a resistor of 1 unit resistance. Thus, V_D ($= I \cdot R$) is **conserved** for resistors in parallel because I and R vary inversely with each other.

Some circuits contain resistors in series and resistors in parallel. Calculations and concepts are straightforward if you understand the fundamentals of single series and single parallel circuit systems. To avoid confusion, think in terms of individual components first.

Remember that EMF (circuit voltage) is consumed entirely; current is **conserved,** i.e., current travels through resistors in series untouched (i.e., current . . . what goes in is what comes out), whereas current divides at a junction, a point in an inverse ratio to the resistances of the resistors contained within that particular parallel connection, and then recombines on emerging from the junction in its totality (i.e., current . . . what goes into a junction comes out); the voltage drop across any resistor can be determined by using Ohm's law: $V = I \cdot R$ (use I_T and R_x values for a

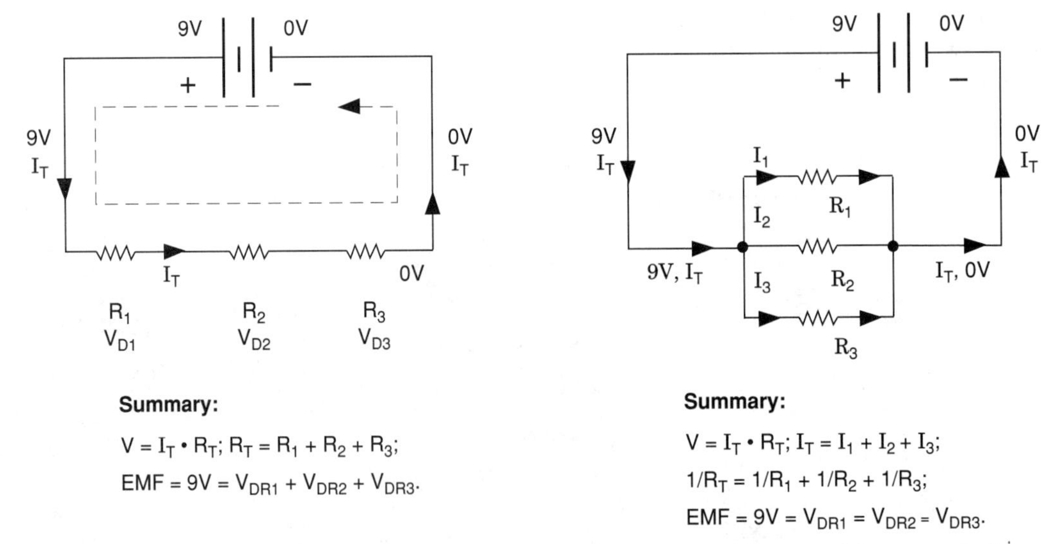

Figure 16-6. A circuit with multiple series resistors (left) and a circuit with multiple parallel resistors (right).

TABLE 16-1. Series versus Parallel Circuit Variables

	Series	Parallel
Resistance (R) for 3 resistors in:	$R_T = R_1 + R_2 + R_3$	$1/R_T = 1/R_1 + 1/R_2 + 1/R_3$
Capacitance (C) for 3 capacitors in:	$1/C_T = 1/C_1 + 1/C_2 + 1/C_3$	$C_T = C_1 + C_2 + C_3$
Voltage drop (V_D) across 3 resistors in:	$V_{DT} = V_{D1} + V_{D2} + V_{D3}$	$V_{DT} = V_{D1} = V_{D2} = V_{D3}$
Current (I) through resistors in:	**Same** I runs through each.	I runs through resistors in an **inverse** ratio of their R

resistor "x" in a series connection, and I_x and R_x for a resistor in a parallel circuit in which I_x is the amount of current travelling through the branch of the junction containing resistor x).

For a circuit containing a single resistor (R_1) in series with two resistors in parallel with each other (R_2 and R_3), calculate the voltage drop across R_1 ($V_{D\,R1} = I_T \cdot R_1$) and subtract this value from the EMF. This calculation gives the remaining voltage that **must** be consumed within the parallel connection ($EMF - V_{D\,R1}$). The voltage drops across both R_2 and R_3 are equal and must be equal to ($EMF - V_{D\,R1}$) because the voltage drop across **all** resistors in parallel is the same (assuming each parallel junction in question contains one resistor).

To determine the total capacitance (C_T) for capacitors in series, use the reciprocal relationship similar to that used for resistors in parallel. For capacitors in a series connection, solve for C_T using the expression: $1/C_T = 1/C_1 + 1/C_2 + 1/C_3$. For capacitors in parallel, use: $C_T = C_1 + C_2 + C_3$. Table 16-1 summarizes series versus parallel circuit variables.

SECTION III

Physics of Waves, Sound, Light, and Nuclear Structure

ph

sics

phys

wave

Wave Characteristics

characteris

Waves are oscillations created by a disturbance. A mechanical wave is set in motion by a disturbance that causes a rapid displacement of a small section of a medium. The particles within the medium do not move with the wave, but create an oscillation so that the wave can move from one place to another. This chapter reviews the concepts and equations that describe both individual waves and repetitive oscillations.

I. Transverse and Longitudinal Motion

When a medium is disturbed, the particles that make up the medium can vibrate back and forth. If a truck tumbles off a bridge into the water below, the truck entering the water is a disturbance that causes the water molecules to start vibrating to and fro about an equilibrium point. The equilibrium point is the previous horizontal level of the water. Over time, the waves travel outward from the truck in concentric circles. This type of wave motion is known as transverse motion, because the water molecules move up and down, while the wave moves horizontally (Figure 17-1, at left). **The direction of motion of the particles is perpendicular to the direction of motion of the wave, resulting in a transverse wave.** Other examples of transverse waves are waves in a rope fixed at the ends, or electromagnetic waves, such as light.

Longitudinal waves are waves in which the oscillations of the particles are in the same direction as the direction of wave motion. A good example is a child's "slinky" toy (Figure 17-1, at right). A slinky is held along the floor by two children. If one child quickly moves a hand forward and then back, the slinky will respond with a compressed area of the coils, followed by a region of rarefaction (an area of rarefaction has wider spaced coils). These areas will then move toward the other child as a wave. Looking at one point on the slinky as the wave passes, you can see that the movement of the particles in the coils is first forward, then backward. Thus, in longitudinal waves, such as slinky waves and sound waves, waves are areas of alternating compression and rarefaction of the particles, caused by oscillations of the particles parallel to the direction of the wave.

II. Wavelength, Frequency, Velocity, and Amplitude

The diagram of a single cycle of a wave (Figure 17-2) shows that the wave consists of a variable that is changing with respect to time in a **sinusoidal** manner. Think of a string stretched from someone's hand to a wall. If the string is moved up and then down quickly, a pulse is generated that will travel the length of the string. If the string is moved up and down repeatedly, a train of waves will be generated that will travel down the string. The speed of the wave train travelling down the string is the same as the speed of a single pulse. These waves have a sinusoidal appear-

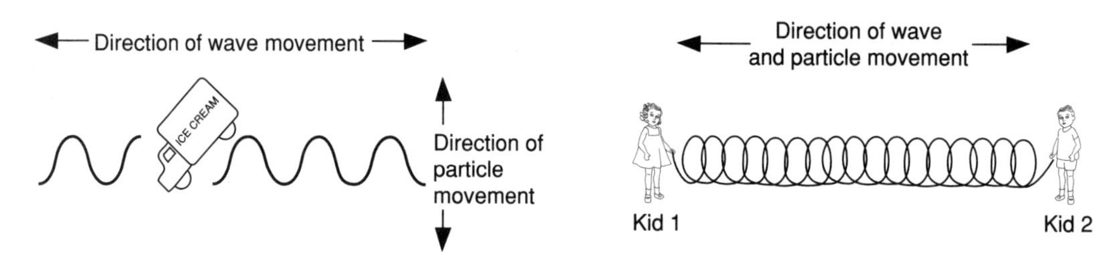

Figure 17-1. Transverse **(A)** and longitudinal **(B)** waves.

ance. Figure 17-2 (at right) also presents a graphic display of the displacement of a certain point on the string versus time.

The **frequency** is how many times per second a complete wave (cycle) passes at a certain point on the string (i.e., **the number of oscillations per unit time**). The frequency in the example in Figure 17-2 is 1 cycle per second.

The **amplitude** of the wave is the maximum displacement of the string. **Amplitude (A) in general refers to the magnitude of maximum displacement in the *y* direction.**

The **velocity (v)** is how fast the wave travels along the string. **The wavelength (λ) is the distance between two points that occupy the same relative position on a wave.** A wavelength is simply the length of one complete wave. In the example in Figure 17-2, the wavelength is the distance between two points on the string that have the same displacement and are moving in the same direction.

The period (T) of a wave is the time (seconds) required for one complete oscillation to pass a given point. Period is the inverse of frequency, which is measured in "per seconds," known as Hertz (1 cycle per second = 1 Hertz).

An important and particularly useful formula that relates frequency, velocity, and wavelength for all waves is:

$$\mathbf{v = f\lambda}$$

If this formula is hard to remember, think about the units of the terms in the equation. Velocity is measured in meters per second, f in 1 per second, and wavelength in meters, so m/sec = (1/sec)(m).

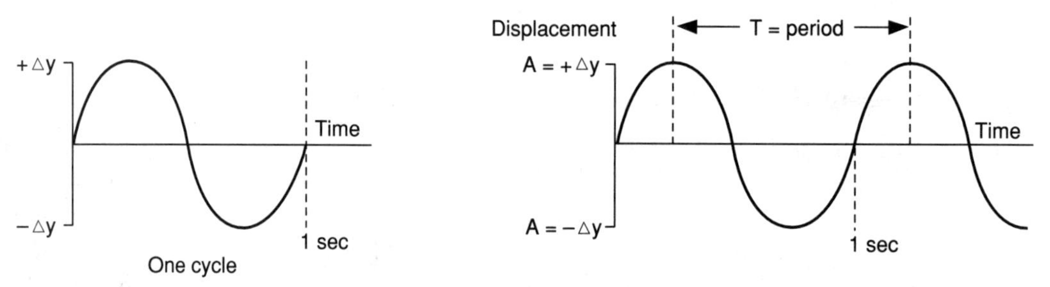

Figure 17-2. At left, a single cycle of a wave; at right, a plot showing the displacement of a single point on a string versus time.

III. Wave Superimposition, Phase, and Interference

If you send a pulse down the string in Figure 17-3, it will travel until it reaches the end of the string, where it will be **reflected.** The reflected pulse will **be inverted when compared to the incoming pulse, because of Newton's third law.** The incoming pulse pushes upward on the end point, causing an equal and opposite reaction force that pushes downward. The reaction force generates the reflected pulse, inverted because of the opposite orientation of the reaction force.

If you send one pulse down from the held end of the string, and another pulse from the fixed end point, the pulses superpose when they meet in the middle of the string (Figure 17-3, at right). If the pulses have the same orientation, they interfere **constructively,** resulting in a **momentary wave of greater amplitude.**

If you send a pulse down the string, allow it to be reflected, and then send another pulse down the string to meet it as it returns, the two opposed pulses would interfere **destructively,** canceling each other.

Note that the amplitude of superposed waves is the sum of the amplitude of the individual waves; this is called the **principle of superposition:**

Amplitude of superposed wave	=	**Amplitude of wave 1**	+	**Amplitude of wave 2**
(y)	=	**(y_1)**	+	**(y_2)**

Recall that the amplitude of a wave on a string is the maximum displacement of the string. Examining displacement as a function of time reveals it is a sinusoidal function. Remember the equation that describes the sinusoidal relationship:

$$\textbf{Displacement} = A \sin(2\pi f t + \phi)$$

This equation allows the displacement to vary sinusoidally from the maximum positive displacement (A) and the maximum negative displacement (−A). Phi (ϕ) is the phase of the wave,

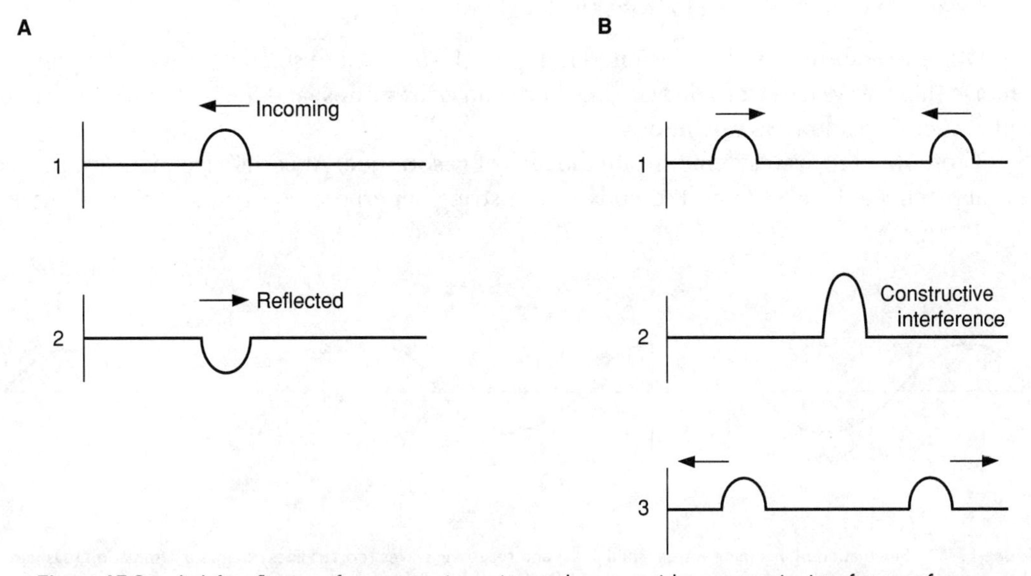

Figure 17-3. At left, reflection of a wave causing an inverted wave; at right, constructive interference of waves.

a term to allow for the point to have an initial displacement at zero time. If the phase of the wave equals 90°, then the displacement at t = 0 is:

$$= \mathbf{A \sin (2 \pi f t + \phi)}$$
$$= \mathbf{A \sin (\phi) = A \sin (90°)}$$
$$= \mathbf{A}$$

Note that although these concepts are related mainly to waves in strings in this discussion, they are equally applicable to other types of waves.

IV. Resonance, Standing Waves, and Nodes

Consider what would happen if you send a continuous train of waves down a string and let them interfere with their return waves. For most frequencies, the interference would be destructive, and no waves would be noticeable in the string. At certain frequencies, however, the incoming and return waves would interfere constructively. The interference effects produced by waves depend on their phases. If two waves reaching a point have their maxima at the same time, they are in phase and add constructively. If a maximum of one wave coincides with a minimum of the other, they are a half wavelength out of phase and interfere destructively. Figure 17-4 shows the constructive and destructive interference occurring between incoming and returning waves. (Note destructive interference illustrated in the middle diagram of Figure 17-4.)

Standing waves result from the interference of two sine waves that have the same frequency and amplitude but are moving in opposite directions (Figure 17-5). Standing waves are observed when a string is vibrating so rapidly that the travelling waves are no longer observed and the string appears to have nonmoving waves.

To examine these ideas in more detail, think about a string tied at both ends, similar to a guitar or piano string (Figure 17-6). When the string is struck, it forms a standing waveform.

Question: What is the standing waveform with the longest wavelength?
Answer: The ends of the string are fixed and cannot move. The middle of the string can vibrate at maximum amplitude, which looks like one half of a sine wave. The wavelength of this waveform would be equal to twice the length of the string.

Other waveforms are also illustrated in Figure 17-6. In these standing waves, the points on the wave that always have zero displacement are known as **nodes,** and the points with maximum displacement are known as **antinodes.**

These standing waves demonstrate the idea of **resonance.** At certain frequencies, the incident and reflected waves from the ends of the string superpose constructively, resulting in a

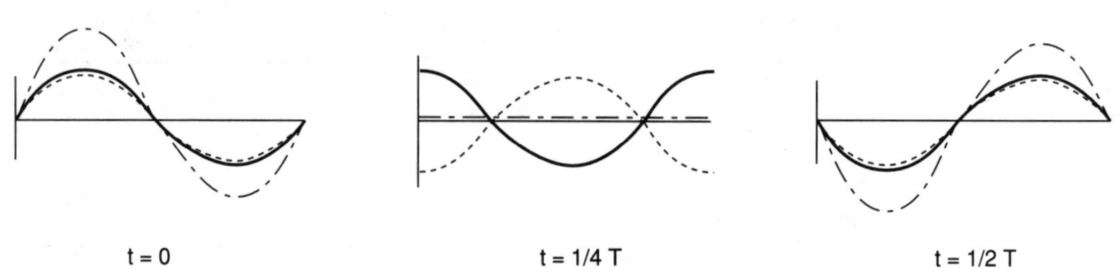

Figure 17-4. Summation of incoming waves (solid lines) and returning waves (dotted lines) to give a summation (dashed lines).

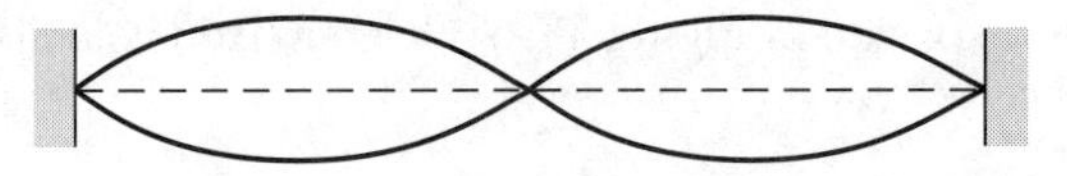

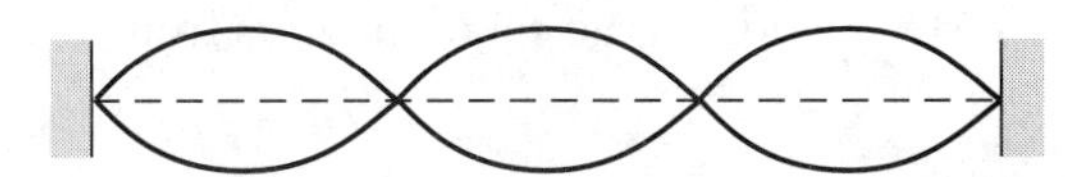

Figure 17-5. Standing waves seen in a string fixed at each end.

standing wave of large amplitude. When one of these standing waves is formed, the waves on the string are occurring at a resonance frequency of the string.

The best way to understand resonance is to think about forces. Energy is supplied most effectively to an oscillating system when an external force acts at the correct frequency. This correct frequency depends on the oscillating system. The optimal frequency causes the greatest amplitude and is known as the **resonant frequency.** If a boy on a swing has a friend apply force (pushing) from behind at just the right interval (correct frequency), the boy can attain the maximum height (amplitude) in the swing. The optimal frequency causing the maximum amplitude of the boy in the swing is the resonant frequency of this system.

The formula $\mathbf{v = \lambda f}$ is important because **it describes the resonance frequency of the wave on the string,** associated with a wavelength and the velocity. **The velocity of waves along a string** can be described by the formula:

$$\mathbf{v} = \sqrt{\frac{\mathbf{F}}{\mathbf{m/l}}}$$

in which v is the velocity of the wave along the string, F is the force of tension on the string, and m/l is the mass of the string per unit length.

This formula is useful when considering a guitar's strings. Some guitars strings are thick and massive per unit length, whereas others are thin. As the force of tension on the string increases, the velocity increases; as the mass/length increases, the velocity decreases. The wavelength of the

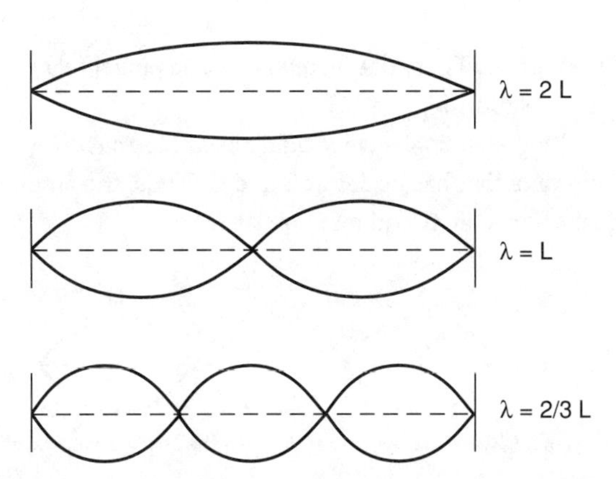

Figure 17-6. Strings tied at both ends, similar to guitar or piano strings. When these strings are struck, standing waveforms result.

possible standing waves is displayed in Figure 17-6. Given a fixed velocity on the string v, the resonance frequencies are:

$$f = \frac{nv}{2l}$$

in which l is the length of the string and n is the set of integers from 1 to infinity.

Example 17-1

TO TEST HOW standing waves on a string work, take a string that is pinned down at the ends and pinch the string at B (a ruler is given beneath for reference):

A B C D string

ruler

1. Would the string have a lower frequency standing wave if it were plucked at A or C?
2. If the string were plucked at A, what would be the amplitude of the vibration at C?
3. If the string were plucked at A, what would be the amplitude of the vibration at D?

SOLUTION:

1. The frequency would be lower when the string is plucked at C, because the wavelength would be longer. Note that v is given by the formula

$$v = \sqrt{\frac{F}{m/l}}$$

so it is constant. Thus, the frequency varies inversely with wavelength, according to the relationship $v = \lambda f$.

2 and 3. The resulting standing wave would resemble that shown in Figure 17-7, with a standing wave that has nodes at B and C. Thus, the amplitude of vibration would be equal to zero at C and maximal at D.

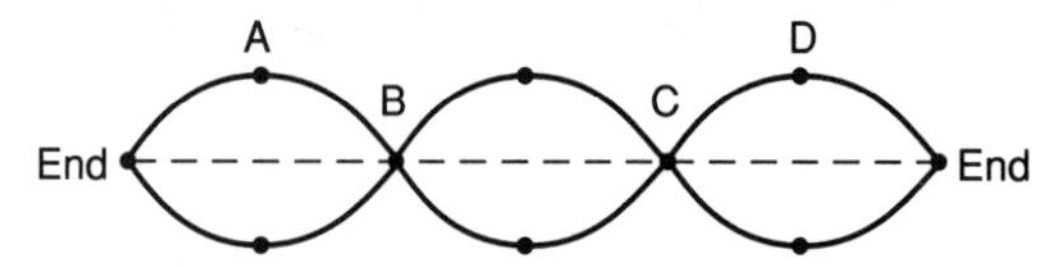

Figure 17-7. The standing wave that would result when the string is plucked at point A in Example 17-1.

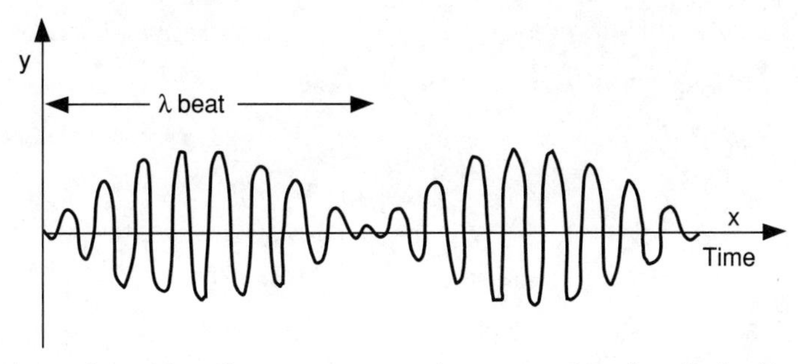

Figure 17-8. Resultant wave formed from the superimposition of two waves of equal amplitudes. Note that the resultant wave has a regular fluctuation in this graphic representation of beat frequency.

V. Beats and Beat Frequency

If you add two travelling waves with slightly different wavelengths and frequencies, it is likely that the waves will add constructively in some regions and destructively in other regions. The resultant wave is likely a rapid oscillation that changes amplitude with time (Figure 17-8). You would expect to find places where the waves add to give a minimum or node as well as places where the waves would add to give maximum amplitudes. This phenomena is called **beating.**

The frequency at which the nodes of the resultant wave pass a given point on the x-axis is the beat frequency. A slightly different definition of beat frequency is **the frequency of the regular fluctuation of two superimposed waves of equal amplitude:**

$$f_{beat} = |f_1 - f_2|$$

circuits

AC Circuits 18

ac circui

This chapter provides a review of the properties of alternating current (AC) circuits, in which the voltage and current vary in a sinusoidal manner, as well as the concepts of inductance, reactance, and impedance.

I. Alternating Current

In Figure 18-1, the currents and voltages of AC and DC circuits are graphically displayed versus time. In DC circuits, the voltage and current remain constant with respect to time. In **AC circuits, the voltage and current are always changing.** In terms of the actual movement of electrons within the wire, the electrons move in one direction, and then the other, driven by the continually changing electrical potential.

Because the current and voltage in the AC circuit are constantly changing, it is useful to have some terms to describe them that do not vary with time, such as I_{max} and V_{max}, which are the maximum current and voltage that occur (see Figure 18-1). Other important terms are **the Root-Mean-Squared (rms) current and voltage,** which define the "**effective**" current and voltage, and are a way of averaging the peaks and valleys. V_{rms} and I_{rms} are determined by the formulas:

$$\mathbf{V_{rms} = V_{max}/\sqrt{2} \text{ or } V_{rms} = 0.71\ V_{max} \qquad I_{rms} = I_{max}/\sqrt{2} \text{ or } I_{rms} = 0.71\ I_{max}}$$

II. Capacitive and Inductive Reactance

Resistors, capacitors, and coils resist the flow of current through an AC circuit. A coil, also known as an **inductor,** consists of many loops of a wire. The opposition to current flow through coils and capacitors is termed **reactance (X),** to differentiate it from the resistance of resistors. Because the **reactance of a coil or capacitor in a circuit resists the flow of current,** reactances are also measured in ohms.

Figure 18-2 shows a coil connected to an AC generator. The type of reactance from the coil is inductive, which is given the symbol $\mathbf{X_L}$.

To identify the property of a coil that acts to resist current flow in an AC circuit, recall Faraday's law. The constantly changing current through the wire in the coil leads to a constantly changing magnetic flux in the center of the coil, which sets up an EMF that opposes the change in flux. Thus, it resists the change in the current. Think of energy being required to constantly change the flux in the coil. Work is done by the power supply's changing electric potential against the coil's induced EMF. This work becomes energy stored in the form of the coil's magnetic field. Because of this loss of energy across the coil, the current seems to encounter resistance.

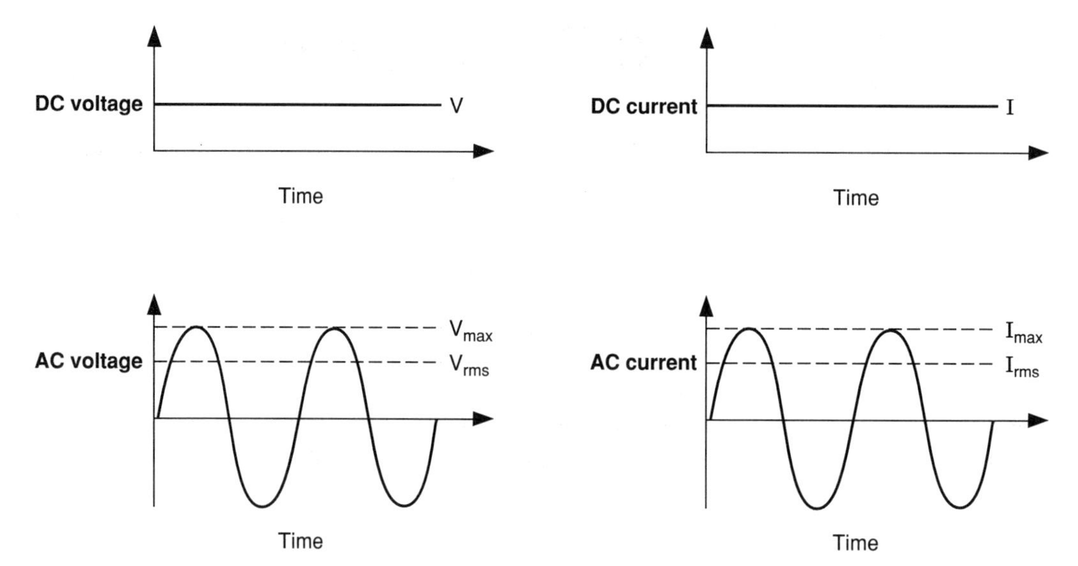

Figure 18-1. Difference between DC and AC voltage and current.

The unit of measurement for a coil's strength (inductance) is the **Henry (H).** The **inductance** of the coil, given the symbol **L**, is proportional to the number of turns of the coil. The reactance of a coil in an AC circuit is described in terms of the inductance of the coil and the frequency of the current. As the frequency increases, the flux changes faster and faster, so the reactance increases:

$$\mathbf{X_L = 2\,\pi\,f\,L}$$

Capacitive reactance (X_C), on the other hand, is caused by the charge storage features of capacitors. Assume that the capacitor in the circuit in Figure 18-3 has a low capacitance, i.e., for a given voltage difference across its plates, it holds little charge. When the generator puts voltage across the capacitor's plates, little current flows onto the capacitor for a given voltage; thus, a small capacitor allows little current flow in the circuit. In contrast, a high capacitance capacitor allows much current to flow for a given voltage difference across its plates, allowing more current flow within the circuit. **The capacitive reactance is thus inversely proportional to the capacitance of the capacitor.**

The frequency of the alternating current also affects capacitive reactance. With a low-frequency current, the voltage across the capacitor plates changes slowly with time, so not much current flows onto and off of the capacitor's plates. With a high-frequency current, the voltage changes quickly, allowing larger currents. Capacitive reactance is:

$$\mathbf{X_C = \frac{1}{2\,\pi\,f\,C}}$$

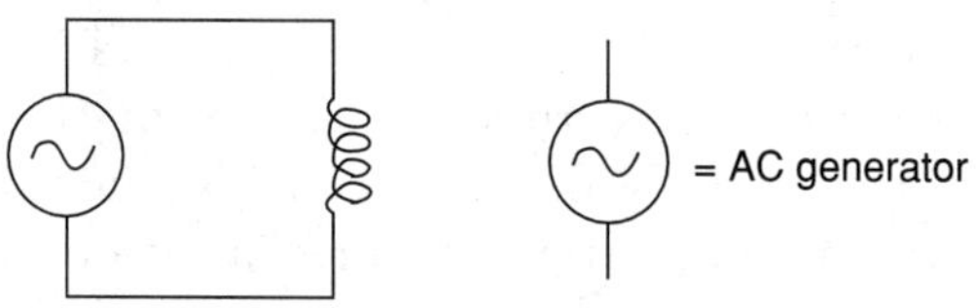

= AC generator

Figure 18-2. An AC circuit containing a coil. This coil has an inductive reactance (X_L).

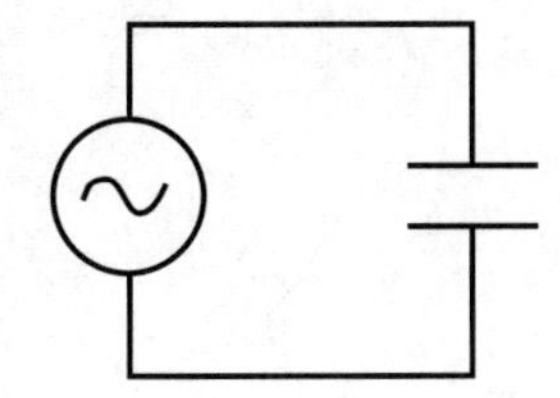

Figure 18-3. A capacitor in an AC circuit.

III. Impedance

Because capacitors, coils, and resistors resist the flow of current within a circuit, their net (effective) resistance when combined within a circuit is **impedance (Z).** The equation for impedance in terms of resistance and reactance is:

$$\mathbf{Z = \sqrt{R^2 + (X_L - X_C)^2}}$$

An important law relating current and voltage in AC and DC circuits is Ohm's law. In DC circuits, recall V = IR. The corresponding formulas for AC circuits are:

$$\mathbf{V_{rms} = (I_{rms})(Z)}$$
$$\mathbf{V_{max} = (I_{max})(Z)}$$

Question: In a given circuit (Figure 18-4), a coil and a capacitor are both connected in series to an AC generator. If the coil has an inductance of 1 Henry, and the capacitor has an inductance of 1 Farad, would there always be a reactance within the circuit?

Answer: The reactance of both the capacitor and the coil depends on the frequency of the generator. Therefore, a more general question would be: At some frequency of the generator, could the inductive and capacitive reactances cancel out? Yes, if the frequency was $\frac{1}{2\pi}$, then $X_L = X_C$, and the reactances would cancel.

IV. Voltage and Current Equations

The function of resistors in AC circuits is similar to that in DC circuits: they resist the flow of current. In Figure 18-5, the voltage and current across a resistor are displayed graphically versus time. Note that the voltage and current always have the same phase.

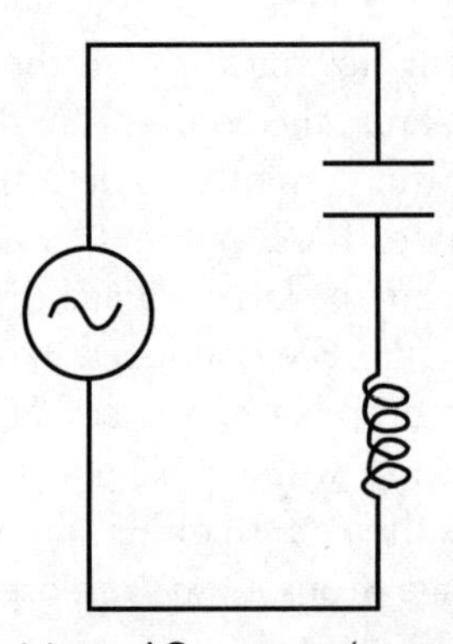

Figure 18-4. A circuit containing an AC generator (power source), a coil, and a capacitor.

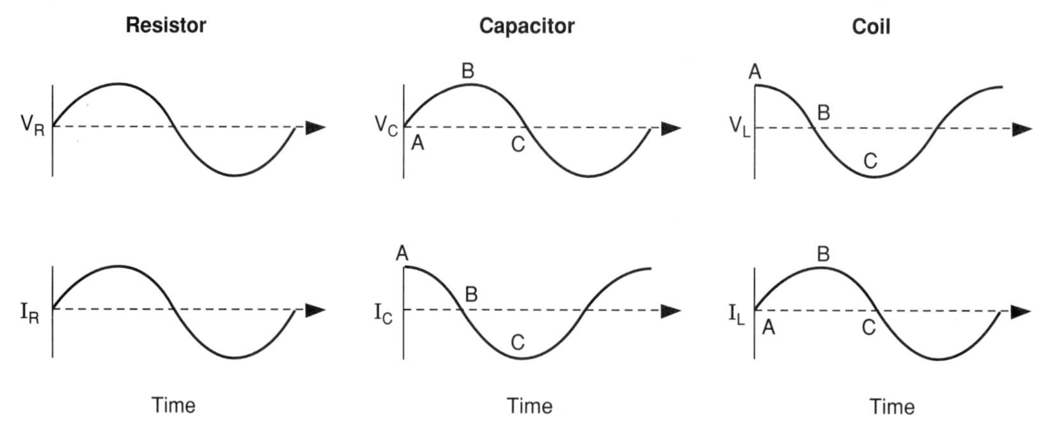

Figure 18-5. Graphs of voltage and current over time for resistors, capacitors, and coils in AC circuits. (See text for detailed explanation.)

Capacitors in DC circuits store charge when a voltage is placed across them. In an AC circuit, they act in the same way. When a capacitor is placed into a circuit with an AC generator, as in Figure 18-3, it experiences the potential difference established by the generator. This sinusoidal potential difference causes charge to flow onto and off of the capacitor.

Look at the AC potential drop (voltage) across the capacitor versus time in Figure 18-5. Note that at $t = 0$, point A, the voltage across the capacitor is increasing rapidly, causing a large current to flow onto the capacitor's plates. At point B, the potential drop across the capacitor has peaked, signifying the largest possible potential pushing charge onto the capacitor's plates. At point B, therefore, the capacitor stops charging, and I drops to zero. At point C, the voltage is sloping downward, meaning less potential pushing charge onto the capacitor's plates. Less charge causes the capacitor to discharge, sending the current in the opposite direction and leading to a negative I in the graph. Continuing to examine the movement of charges yields the two sine waves for V and I in Figure 18-5.

Looking at the sinusoidal curves of V and I across the capacitor in the AC circuit reveals an important difference from a resistor AC circuit. In the resistor circuit, V and I vary in the same way with respect to time. In the capacitor circuit, the I sine wave leads the V sine wave by 90°; i.e., a phase difference of 90° exists between the two graphs. Thus, I_{max} precedes V_{max} by 90°.

In an AC circuit with a coil (inductor), the coil's properties also result in phase shifting between current and the potential drop across the coil versus time. Remember that the coil requires energy to constantly change the flux of its magnetic field, and this energy loss is seen as a voltage drop across the coil. Connecting the coil to an AC generator yields the V_L versus time and I_L versus time graphs in Figure 18-5.

As with the capacitor, compare the potential drop across the coil to the current flowing through the coil. Examine the current flowing through the coil, and use that knowledge plus Lenz's law to find the potential drop across the coil. At A in the I_L versus time graph, the current through the coil is zero, but it is increasing rapidly. To determine what will happen to the voltage drop, remember Lenz's law, which states $EMF = -\Delta\phi/\Delta t$. When the current increases rapidly, the flux will also increase rapidly, causing the large voltage drop across the coil seen in Figure 18-5. As the current stabilizes at the peak, the EMF induced in the coil falls to zero, because the flux is no longer changing, i.e., no voltage drop across the coil at this point. As the current moves in the opposite direction to point C, an EMF will oppose the change in flux, although the EMF will cause a negative potential drop across the coil. In effect, the coil is releasing the energy stored as the magnetic flux. Note that the voltage drops always oppose the changes in flux; thus, they lead the current wave by 90°.

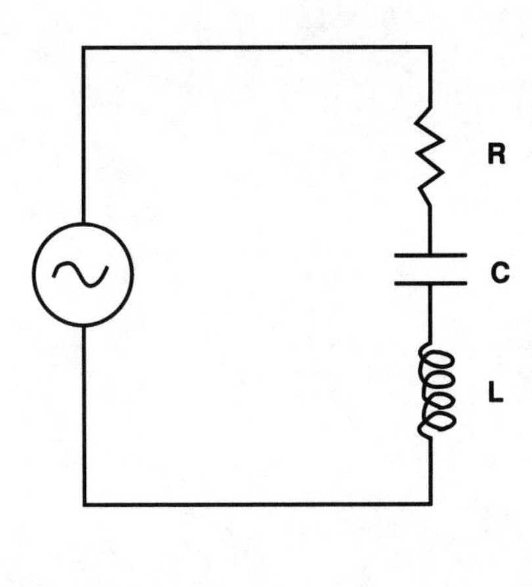

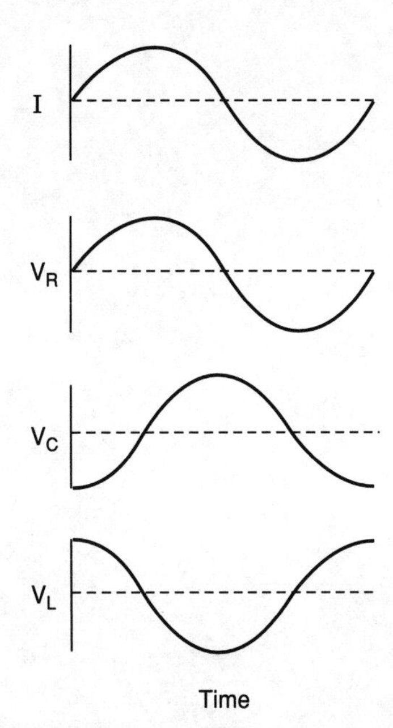

Figure 18-6. An AC circuit containing a resistor, a capacitor, and a coil (left). A plot of current, and the voltage drops across each element in the circuit (right). Note the relationship between the voltage drop and current in each case. (See text for more discussion.)

Using a circuit such as that shown in Figure 18-6, you can show the voltage drops across each element of the circuit. The voltage drop of the resistor is always in phase with the current; as more current flows through the resistor, more heat dissipates and the voltage drop increases. The voltage drop across the capacitor follows the current by 90°; the capacitor has an increasing potential difference as charge flows onto it, and a decreasing potential difference as charge flows off. The voltage drop across the coil leads by 90°; Lenz's law states that the induced EMF (the potential difference across the coil) always opposes the change in flux versus time.

The final concept to understand is the concept of power dissipation in an AC circuit. Power is dissipated in a resistor ($P = I^2R$), but no power is dissipated by capacitive and inductive circuit elements.

simple harmonic motion 19

Simple Harmonic Motion

In simple harmonic motion, displacement varies in a sinusoidal manner with respect to time. The types of simple harmonic motion reviewed in this chapter involve pendulums and masses attached to springs. Concepts involving the energy of an oscillating system are also addressed.

I. Periodic Motion and Hooke's Law

In **simple harmonic motion,** a disturbance causes an object to vibrate about an equilibrium point. Figure 19-1 includes a graph of the displacement of the object versus time (sine wave) and the formula used to find the displacement of the object at any given time.

The formula in Figure 19-1 shows that displacement varies as a sine wave, multiplied by A, the maximum amplitude of the displacement. A pendulum swinging back and forth rapidly will have a high frequency, so $2\pi ft$ will increase rapidly with increasing time, and the displacement will move rapidly from A to $-A$. The frequency (f) is expressed in terms of cycles per second. Remember from previous discussion of circular motion that $2\pi f$ is ω, the angular frequency in radians per second, giving:

$$\textbf{Displacement} = \textbf{A} \sin (\omega\, \textbf{t})$$

SPRINGS

In Figure 19-2, a spring is attached at one end to a wall and at the other end to a mass. The mass is resting on a frictionless surface at $\Delta X = 0$. Consider what would happen if you disturb the system. One way to disturb the system involves pulling the mass outward to the point $\Delta X = X_{max}$, but how much force would be required? The amount of force required depends on the properties of the spring; for example, it would be harder to pull the mass with a thick spring. It also depends on how far you stretch the spring. It makes sense that the farther you stretch the spring, the harder you will have to pull.

This reasoning leads to the following equation, called Hooke's Law:

$$\mathbf{F = -k\, \Delta x}$$

which states that the force required will be equal to the displacement times k, the spring constant. The spring constant (k) reveals the stiffness of the spring. The right side of the equation is negative because Hooke's law actually tells how much force the spring will exert **against** any pull, so it is always directed against the direction of the displacement. A positive displacement will cause a negatively directed force from the spring. The force of the spring always being proportional to

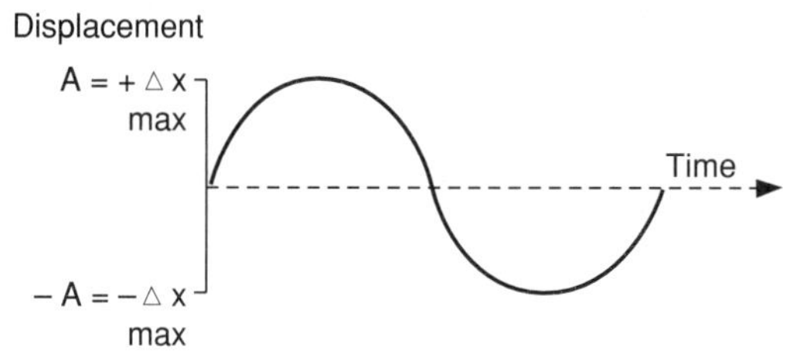

Figure 19-1. Displacement of an object in simple harmonic motion described by formula and graphically. Displacement = A sin (2πft).

the displacement and directed opposite to the displacement is what defines simple harmonic motion. Using Newton's second law, F = m a, the acceleration of the mass will be:

$$a = -k\,\Delta x/m$$

Another formula that remains to be examined in the mass spring system involves what Hooke's law (F = $-k\,\Delta$ x) can tell you about the frequency of the motion of the mass. Thinking intuitively, a strong spring will exert large forces on displacement, and it likely will pull the mass back fast, causing a higher frequency of vibration with a larger spring constant, k. If the mass is large, the acceleration caused by the force of a given spring would be relatively small, so you could guess that the frequency would decrease with increasing mass. Actually, when derived, the frequency and period can vary with the square roots of k and m. **The following formulas detailing the relationships between frequency, period, m, and k are important.** Remember that period (denoted T) = 1/f.

$$f = \frac{1}{2\pi}\sqrt{k/m}$$

OR

$$T = 2\pi\sqrt{m/k}$$

SIMPLE PENDULUM

Another example of simple harmonic motion is the simple pendulum, in which a mass is suspended from a massless rod. When the pendulum swings back and forth with an angle of displacement $\theta < 10°$, it follows simple harmonic motion: the restoring force on the pendulum is proportional to the pendulum's angle of displacement. **The motion of the pendulum is described by the formulas:**

$$f = \frac{1}{2\pi}\sqrt{g/L}$$

$f = \frac{1}{T}$ or

$T = \frac{1}{f}$

$$T = 2\pi\sqrt{L/g}$$

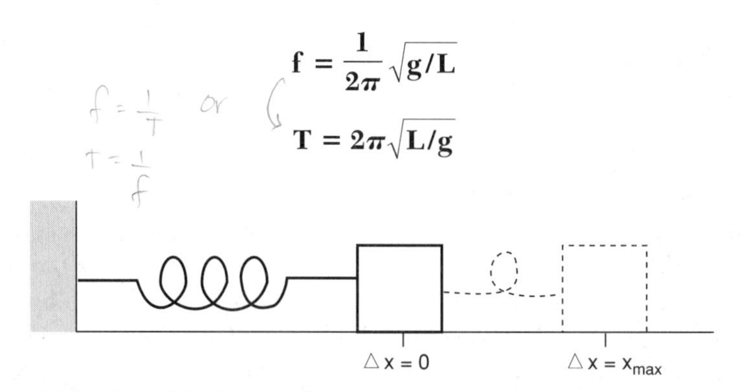

Figure 19-2. A simple harmonic oscillating system: a spring attached to a mass and to a wall.

EXAMPLE 19-1

A PENDULUM WITH A 5-kg ball and a length of 10 m is displaced by 10°.

1. What is the frequency of its simple harmonic motion?
2. If the mass is doubled to 10 kg, what is the period of its simple harmonic motion?

SOLUTION:

1. Use the formula just provided to derive the frequency of the pendulum. L is given as 10 m; g was not given, and in such cases, it is assumed to be earth's gravity of 10 m/sec^2. Thus,

$$\mathbf{f = \frac{1}{2\pi}\sqrt{10/10}}$$

$= 1/2\pi$ cycles per second (Hz), **answer.**

2. The frequency and period of the pendulum do not depend on the mass of the pendulum. If you put a heavy mass on the end of the pendulum, gravity will pull it more strongly, but it will have a greater mass to accelerate. These effects cancel, causing the frequency to be independent of mass. To find the period, use the formula: $\mathbf{T = 2\pi}\sqrt{L/g}$; thus, $T = 2\pi$ seconds, answer.

Note: More information was given than was required for solving the problem. It is important to be able to sort out what information given is required to solve the problem, and what information is unnecessary.

II. Kinetic Energy and Potential Energy of an Oscillating System

The kinetic energy (KE) of an oscillating system is expressed using the formula:

$$\mathbf{KE = 1/2\ mv^2}$$

Because $v_{max} = \omega A$, the maximum kinetic energy of the oscillating system in terms of its amplitude and angular frequency is:

$$\mathbf{KE_{max} = 1/2\ m\omega^2A^2}$$

The potential energy (PE) of the oscillating mass spring system is:

$$\mathbf{PE = 1/2\ k\ \Delta\ x^2}$$

in which Δ x is the displacement from equilibrium. (Usually, x is defined as being = 0).

Example 19-2

A MASS OF 4 kg resting on a frictionless surface is attached to a massless spring of spring constant k=10 newtons per meter at rest. A ball of tape of 1 kg moving at 10 meters per second collides with the mass spring system and sticks. Using the diagram in Figure 19-3, determine the resulting amplitude of the simple harmonic motion.

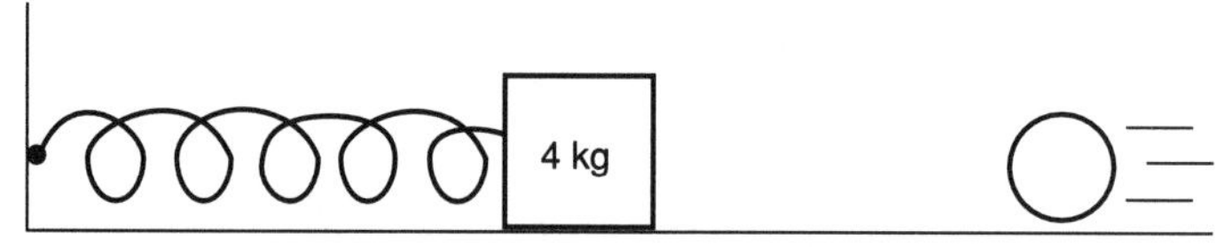

Figure 19-3. A 1-kg tape ball moving at 10 m/sec collides with a mass spring system (see Example 19-2 for details).

SOLUTION:

This problem addresses two main concepts. The first part of the problem concerns the collision between the ball of tape and the mass attached to the spring. Because the ball of tape hits and sticks, it is a completely inelastic collision. During an inelastic collision, only conservation of momentum applies, because much energy is lost to deformation and heat. Considering conservation of momentum:

$$m_1 v_1 = (m_1 + m_2) v_f$$

$$v_f = \frac{m1v1}{m1 + m2} \qquad v_f = \frac{(1)(10)}{1 + 4}$$

$$v_f = 2 \text{ m/sec, answer.}$$

Knowing the velocity of the tape ball and mass immediately after the collision, examine the conversion of kinetic to potential energy as the spring is pushed back toward the wall. Considering conservation of energy:

$$\mathbf{KE = PE}$$

$$\mathbf{1/2\ m_{total}\ v_f^2 = 1/2\ k\ \Delta x^2}$$

$$\mathbf{1/2 \cdot 5 \cdot 2^2 = 1/2 \cdot 10 \cdot \Delta x^2}$$

$$\mathbf{\Delta x = \sqrt{2}\ m}$$

This problem is analogous to the "ballistic pendulum," in which a bullet strikes a hanging block. The same steps are taken in looking for the maximum height to which the pendulum + bullet rise: $PE = m_{(total)}gh$. If h is small compared to l (thus giving a small θ), then the resulting motion would be simple harmonic motion.

Sound 20

Simple vibrations of matter are responsible for the variety of sounds we hear, from the minute buzz of a mosquito's wings to the loud sounds of a rock concert. Sound can be transmitted through many types of matter. This chapter is a review of how sound travels in air, liquids and solids.

I. Basic Concepts

A. WHAT IS SOUND?

Sound waves are longitudinal waves that can be transmitted through solids, liquids, or gases. Because they are longitudinal waves, the molecules vibrate in the **same direction** as the direction in which the wave travels. The frequency of the oscillations can vary greatly, giving rise to the wide variety of pitch in the sounds perceived. Perceptible sounds vary from 20 to 20,000 Hz. Sound waves are set up by a vibrating object or air column, and travel outward in three dimensions from the source.

B. SPEED OF SOUND IN SOLIDS, LIQUIDS, AND GASES

WHAT DETERMINES THE SPEED OF SOUND IN SOLIDS AND LIQUIDS?

Think about a sound wave, which is **a pressure pulse travelling through a material.** A material has an area of compression and rarefaction that travels through it at a certain velocity. This area of compression and rarefaction is caused by the molecules first moving forward, then re-equilibrating by moving backward.

A factor that might affect the speed of sound is the stiffness of the material—its resistance to compression, also known as the bulk modulus, $\Delta V/V$. A stiff material cannot be compressed easily, which causes the sound wave to travel faster. Increasing the density of the material has the opposite effect. The pressure pulse of the sound wave must accelerate the substance, first forward and then backward, as it travels. **With a denser substance, these accelerations are slower.** Consider the quantitative formula:

$$\mathbf{v} = \sqrt{\mathbf{BM}/\rho}$$

which shows that the **velocity of a sound wave in a solid or liquid is proportional to the square root of the bulk modulus (BM) divided by the density.**

In air, the same ideas determine the speed of sound. One constant worth remembering is the speed of sound in air of normal composition at room temperature, 344 m/sec.

WHAT HAPPENS TO THE SPEED OF SOUND IN AIR AS THE TEMPERATURE INCREASES?

As temperature increases, the thermal motion of the molecules also increases, allowing the pressure pulse to travel faster. The velocity of sound in a gas is proportional to the square root of the absolute temperature in Kelvins.

II. Sound Intensity, Pitch, and the Decibel

Pitch is the perceived predominant frequency of a sound source. In the audible range, frequencies vary from 20 to 20000 Hz. Sounds with a frequency lower than 20 Hz are **infrasonic waves;** sounds with frequencies greater than 20000 Hz are **ultrasonic waves.** Note that in air, the wavelength of sound varies with frequency, $\lambda = v/f$, as the waves travel at a constant 344 m/sec.

When you hear a sound from a sound source, you are perceiving the pressure pulses. These pulses have energy that can be measured to determine their intensity. With a continuous sound, pressure pulses continue hitting with time, so they have power (remember P = energy/time). **Sound intensity is the amount of power (P) per unit area (A):**

$$I = P/A$$

Sound intensity falls off as $1/r^2$, in which r is the distance from the sound source. This relationship is similar to gravitational and electric fields, all of which expand in three dimensions. **The critical relationship between sound intensity and distance from the sound source is important to remember.**

One way to understand sound intensity is to examine the differences between the intensities of certain sounds. The quietest sound that "good" ears can detect has a sound intensity of 10^{-12} watts/m^2. A quiet room has a sound intensity of approximately 10^{-9} watts/m^2. Normal conversation in a room produces a sound intensity of 10^{-6} watts/m^2. The front row of a heavy metal concert, on the other hand, has a sound intensity of approximately 1 watt/m^2.

It is difficult to portray such huge differences in sound intensity or even to think about a scale in which the numbers can vary by a factor of 10^{12} or more. To make the differences between sounds easier to write, and to provide a means to accurately describe the perceived loudness of sounds, the decibel scale was created. **The decibel scale is a logarithmic scale** in which zero dB was set equal to the quietest possible sound intensity, 10^{-12} watts/m^2. All other decibel levels were measured in terms of the reference point I_0, according to the formula:

$$dB = 10 \log (I/I_0) \text{ in which } I_0 = 10^{-12} \text{ watts/m}^2$$

EXAMPLE 20-1

A SOUND SOURCE MOVES from 2 m to 4 m away from a sound receiver. Explain how the sound intensity changes with this move.

SOLUTION:

Think of the sound intensity at 2 m as $1/2^2$, or 1/4. After the move, the intensity is $1/4^2$, or 1/16. Thus, when the distance from the source doubles, the sound intensity falls by a factor of 4 (from 1/4 to 1/16).

Example 20-2

THE SOUND INTENSITY at a distance of 1 m from a stereo speaker is 10^{-4} watts/m^2. Determine the approximate sound intensity level (in dB) at a distance of 3 m.

SOLUTION:

Because sound intensity varies as $1/r^2$, the sound intensity at 3 m will be 1/9 that of 1 m. Thus, the intensity level will be approximately equal to $10 \log [(\frac{1}{9} \cdot 10^{-s})/10^{-12}]$, or about 70 dB.

When speaking about sound intensity in the decibel scale, the term **intensity level** is used. Note that the intensity level in dB is not a measurement of intensity, but is a comparison of the intensity of one sound to the intensity of the quietest sound that can be heard. In decibels, a quiet room is about 30 dB, normal conversation is about 60 dB, and a loud rock concert is in the range of 120 dB.

Because decibels are logarithmic measurements, you cannot directly add decibel units. If two people, each speaking at an intensity level of 50 dB, speak simultaneously, the total intensity level is not 100 dB. The addition of the intensity levels of these two speaking voices is about 53 dB. Thinking logically, if 10 voices at 50 dB speak simultaneously, a 10-dB increase in sound intensity level (in the decibel scale) will be heard, because a 10-dB increase corresponds to a 10-fold increase in the I values. Ten voices, each at 50 dB, give a total sound intensity level in decibels of 50 dB + 10 dB = 60 dB. Thus, the sound intensity level of two voices speaking simultaneously will predictably be less than 55 dB.

If you wish to calculate this log relationship, start by converting the decibel values to I values. Add the two I values and apply the decibel formula. Note the log 2 = 0.3.

III. Doppler Effect

As cars go by during an Indianapolis 500 race, the frequency of the sounds seems to change. In Figure 20-1, one car is approaching a listener and one is receding from a listener. The cars are both emitting sound of the same frequency, but the sound waves of the car approaching the lis-

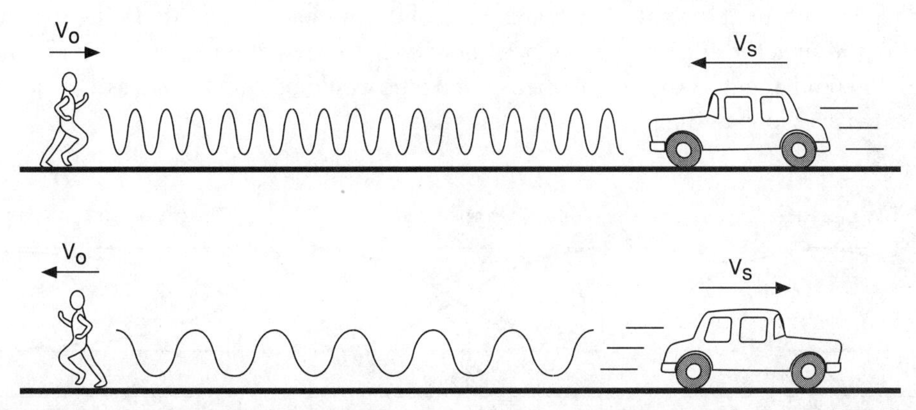

Figure 20-1. A source approaching a listener (top) and a source receding from a listener (bottom) helps to understand the Doppler effect.

tener are compressed slightly because the car is moving closer each time a new wave is emitted. With the receding car, the waves are stretched slightly because the car moves away during each wave.

The compression and stretching of the waves is perceived as a change in frequency, because the sound waves are still travelling at 344 m/sec, yet their wavelength has changed. (Remember $f = v/\lambda$.) An equation that describes the phenomenon follows:

$$f' = f\left(\frac{1 \pm v_o/v}{1 \mp v_s/v}\right)$$

in which $\mathbf{v_o}$ is the velocity of the observer, $\mathbf{v_s}$ is the velocity of the source, and **v** is the speed of sound. The + (numerator) and − (denominator) signs are used if the source is approaching the observer or the observer is approaching the source. The − (numerator) and + (denominator) signs are used if the source is fleeing from the observer or the observer is fleeing from the source.

Remember: the perceived frequency increases if the source or observer is approaching, and the perceived frequency decreases if the source or observer is receding.

IV. Resonance in Pipes and Strings

Chapter 17 provides a review of the introductory concepts of resonance in pipes and strings. Recall that a standing wave was created on a string by waves of certain frequencies undergoing constructive interference with the return waves from the end points.

On a string, nodes are always at the ends, because string movement is confined by being tied. The fundamental wave (lowest frequency, longest wavelength) has an antinode in the exact center. Other waveforms can be drawn by putting additional nodes and anti nodes on the string. Because the velocity is constant, the wavelength and frequency are related by:

$$v = f\lambda$$

Another example of standing waves is the vibration of air within a pipe. Waves on a string are transverse waves; the air within the pipe has longitudinal waves. **Three different types of pipes to remember are: closed at both ends, closed at one end, or open at both ends.**

A standing waveform for a pipe that is closed at both ends (Figure 20-2) is similar to a string's waveform. The nodes at each end are caused by the closed ends; the air is unable to move back and forth because its movement is stopped by the lack of movement of the end. In the center of the pipe, the air at the antinode is able to freely move back and forth. In the pipe with two open ends, the air at the ends is able to move freely, so there are antinodes at each end. The wavelength of the fundamental wave for an open ended pipe would be equal to 2L, as a complete wave would be twice as long as the pipe.

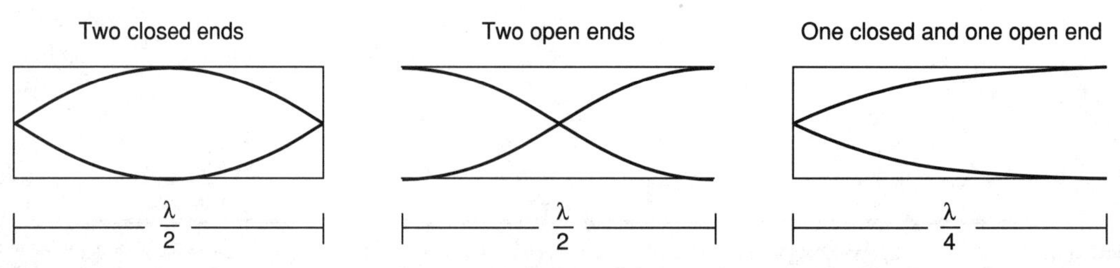

Figure 20-2. Waveforms in pipes with closed ends (left), open ends (center), and one closed end, one open end (right).

The pipe with one closed and one open end has somewhat different properties (see Figure 20-2). Like the other pipes, the closed end has a node and the open end has an antinode. Unlike the other pipes, the wavelength of this fundamental wave would be four times as long as the length of the pipe.

V. Harmonics

Harmonics are resonance frequencies that are multiples of the fundamental frequency, including the fundamental frequency itself. They are numbered consecutively, with the fundamental being the first harmonic.

Figure 20-3 illustrates the harmonic frequencies for each of the types of pipe. The second harmonic wave has an additional node within the pipe; the third harmonic wave has an additional two nodes. The open and closed ends of the pipes have the same nodes and antinodes as the fundamental. The second harmonic is also called the **first overtone** (first tone "over" the fundamental), the third harmonic is also known as the **second overtone,** and so on.

The frequencies associated with the various harmonics with the different types of pipes are as follows. **The "both end closed or open" pipes have the same frequencies as a string, according to the formula:**

$$\mathbf{f = v/\lambda = (v)(n)/2L}$$

The various frequencies are found by substituting for n any positive integer (1,2,3 . . .) and also substituting the length of the pipe, L.

The "open at one end and closed at one end" pipe is slightly different. Use the following formula for its resonance frequencies:

$$\mathbf{f = v/\lambda = (v)(n)/4L}$$

in which n is the set of positive odd integers (1,3,5 . . .). These formulas, which arise from the familiar $f = v/\lambda$ formula, are found by deducing the possible values for the wavelength in terms of L, as shown in Figure 20-3. It is important to understand these formulas, and to be able to draw waveforms for the various pipes.

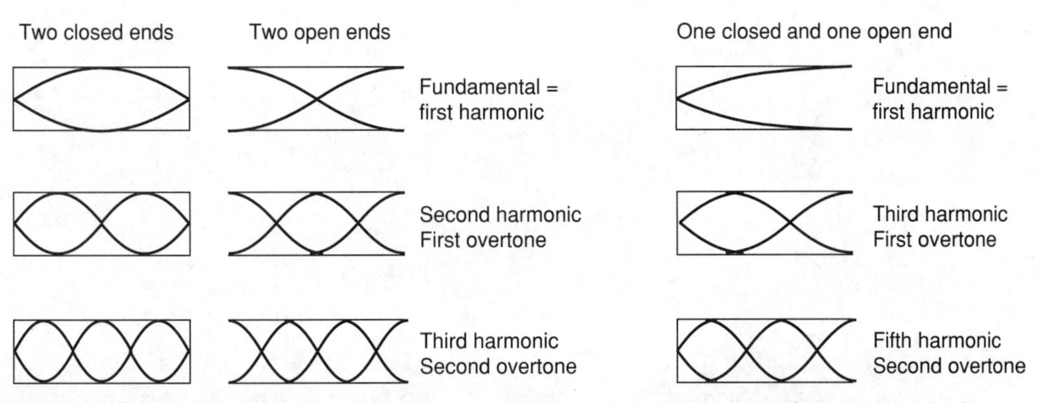

Figure 20-3. The first harmonic frequency (fundamental frequency), and additional harmonic frequencies for the three different pipe types. Note the nodes, antinodes, and the difference between the different pipe types based on their open and closed ends. (See text for explanation.)

Light and Optics

21

This chapter provides a review of electromagnetic waves and the concepts and components of visible light.

I. Electromagnetic Waves and the Visual Spectrum

An **electromagnetic wave** is an electromagnetic disturbance that travels through space at a speed of 3×10^8 meters per second. As shown in Figure 21-1, **the electromagnetic wave is a transverse wave;** the electric and magnetic fields (on the x and y axes) vibrate **perpendicular** to the direction of the propagation of the wave (along the z axis).

Electromagnetic waves have a constant velocity in a vacuum, and their frequency and wavelength are inversely proportional. The properties of electromagnetic waves vary with frequency and wavelength, giving rise to the electromagnetic spectrum. The waves with the longest wavelength are radio waves, commonly used in AM, FM, and television broadcasting. At the shortest wavelength are gamma rays, the most energetic electromagnetic waves. The complete electromagnetic spectrum (Figure 21-2), with an expanded view of visible light, ranges from a wavelength of 700 nm for red to 400 nm for blue.

II. Polarization of Light

Light from most sources is unpolarized, meaning that the direction of the E- and B-fields are at random, even when the light is moving in the same direction (same z axis). In polarized light, the waves have aligned electric and aligned magnetic fields (Figure 21-3).

One of many ways for light to become polarized is for it to pass through a polarizing filter. Polarizing filters are plastic sheets with millions of aligned, long linear molecules. When light passes through the polarizing filter, those light waves with electric fields parallel to the molecules set up currents within the molecules and are thus absorbed. Light waves that have electric fields perpendicular to the molecules are not absorbed.

When two perfectly made polarizing filters are placed at right angles to one another, all of the incident light will be absorbed.

III. Refraction of Light and Snell's Law

When light strikes an interface between two media, part of the light is reflected and part is transmitted. The transmitted light is refracted at the **interface,** meaning that the light waves bend. Refraction leads to several important ideas within physics, and is the underlying mechanism behind the function of lenses and prisms.

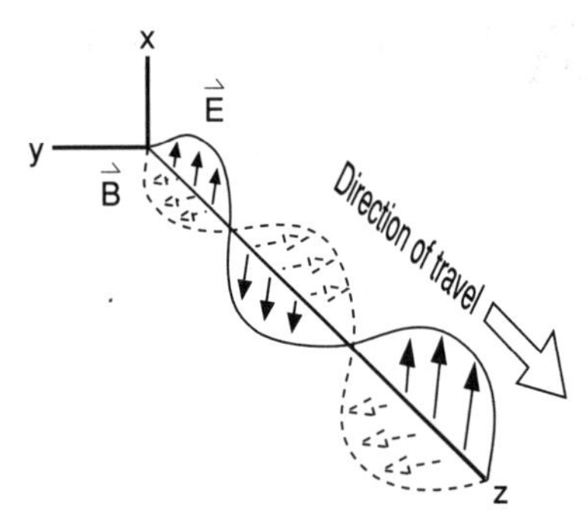

Figure 21-1. The electromagnetic wave is a transverse wave, with electric and magnetic fields oscillating in phase perpendicular to each other. Note the direction of travel of the electromagnetic wave.

	TV/AM	FM	Radar/Microwave	Infrared	Visible	Ultraviolet	X-rays	Gamma rays
f (Hz):	10^6	10^8	10^{10}	10^{12}	10^{14}	10^{16}	10^{18}	10^{20}

	Red	Orange	Yellow	Green	Blue	Violet
f (Hz):	$4 * 10^{14}$	$5 * 10^{14}$		$6 * 10^{14}$		$7 * 10^{14}$
λ (nm):	700	600		500		400

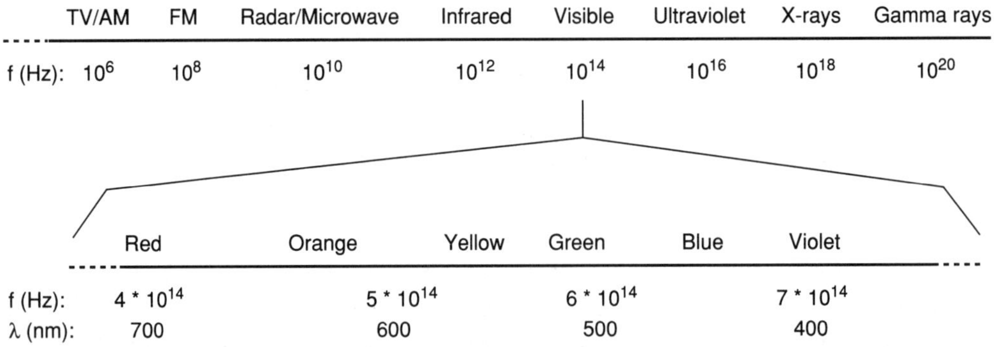

Figure 21-2. The electromagnetic spectrum. Note the expanded view of the visible light frequencies.

When light waves enter a transparent medium other than a vacuum, such as glass, they ***slow***, travelling at less than 3×10^8 m/sec. This concept is the basis for the **index of refraction, n,** of a medium, given by the formula:

$$\mathbf{n = c/v}$$

in which **c** is **3×10^8 m/sec,** the speed of light in a vacuum, and v is the speed of light in the medium where it is slowed. Thus, **for a vacuum, n = 1, and for other media, n > 1.**

If the light wave enters a new medium perpendicular to the surface (Figure 21-4, at left), it will continue in the same direction. Should the light wave enter at an angle, however, it will **bend toward the normal vector (to the surface that it encountered) if the n value of that new medium is greater than the medium from which it came.** (The normal vector is a vector perpendicular to a given surface.)

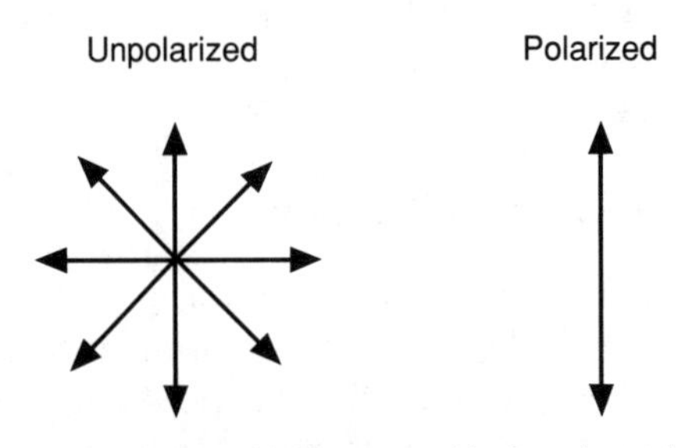

Figure 21-3. Unpolarized versus polarized light as viewed looking down the z-axis at the E-field alone.

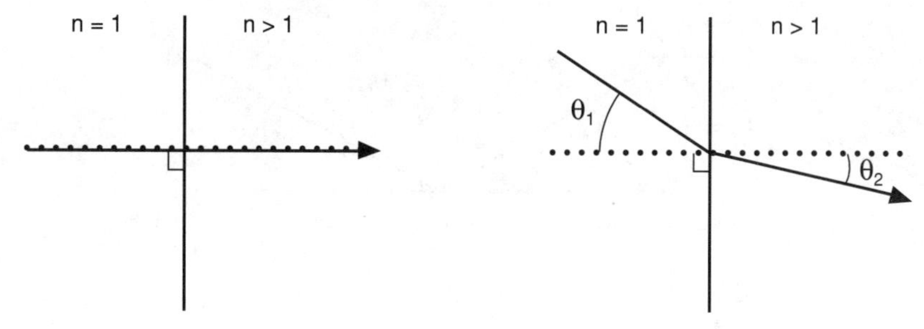

Figure 21-4. Light striking perpendicular to a surface with an index of refraction (n) greater than 1 (at left); light striking an interface with an incident angle, and bending toward the normal when n is greater than 1 (at right).

These concepts should help you understand the diagram in Figure 21-4. The incident ray hitting a surface with $n > 1$ forms an angle θ_1 to the normal vector. The refracted ray forms an angle θ_2 that is less than θ_1.

Note that as v slows, the ray bends toward the normal vector, and if v accelerates, the ray bends away from the normal vector. Thus, when a ray goes from a material with a low index of refraction to a high index of refraction ($n_{incident} < n_{refracted}$), it will bend toward the normal. If the ray goes from a material with a high index of refraction to a low index of refraction material ($n_{incident} > n_{refracted}$), it will bend away from the normal.

Example 21-1

AS A CHILD is about to put a quarter into a machine to buy food for fish in a small pond, he drops the quarter and it rolls into the water. The coin seems to be close to the surface and easy to reach. Explain why the quarter might not be as near the surface as it appears to be.

SOLUTION:

As shown in Figure 21-5, the eyes sense depth by noticing the angle at which they converge in order to focus on an object. Because the quarter is under water, however, rays from the quarter reach the surface and then are bent away from the normal. The eyes interpret the bent rays as showing that the quarter is at depth d_2 instead of d_1.

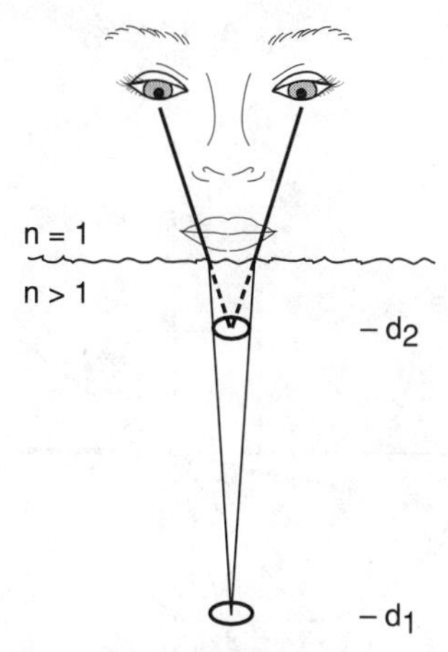

Figure 21-5. Explanation of how a submerged coin may appear close to the surface (see Example 21-1 for details).

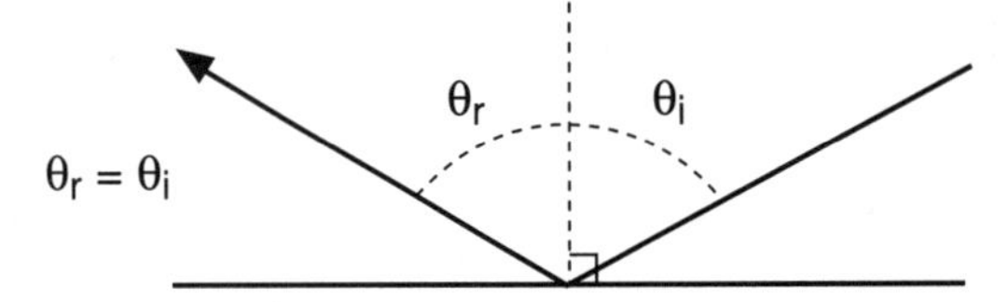

Figure 21-6. Incident light reflects off a surface such that $\theta_i = \theta_r$.

The angles are in all cases the **angles between the rays and the normal vector of the surface.** The exact relationship between θ_1 and θ_2 is expressed by **Snell's law:**

$$\mathbf{n_1 \sin \theta_1 = n_2 \sin \theta_2}$$

or

$$\mathbf{v_2 \sin \theta_1 = v_1 \sin \theta_2 \text{ (since } n = c/v)}$$

IV. Reflection and Total Internal Reflection

Reflection is a familiar concept. You expect light waves to "bounce off" of objects. When a light ray strikes a surface, part or all of the incident light is reflected at an angle θ_r, called the **reflected angle.** Figure 21-6 illustrates the concept of reflection.

The concept of total internal reflection is important. An incident ray from within a substance with $n > 1$ approaches an interface to a substance, air, or vacuum with a smaller n. As shown by the ray 1 in Figure 21-7, the ray will be refracted away from the normal vector. As the angle θ_i increases, you reach ray 2, and the refracted ray starts coming closer and closer to being parallel to the interface. Note that at ray 3, at the incident angle θ_c, the refracted ray actually skims the interface surface. **This angle is the critical angle,** because if θ_i increases still farther, none of the incident light will be refracted, and all will be reflected. This phenomenon is labeled **total internal reflection.** Where $\theta_i = \theta_c$, then $\theta_r = 90°$. Another way to derive the relationship between the indices of refraction and the critical angle follows:

$$\mathbf{n_i \sin \theta_i = n_r \sin \theta_r}$$

$$\mathbf{\sin \theta_i / \sin \theta_r = n_r / n_i}$$

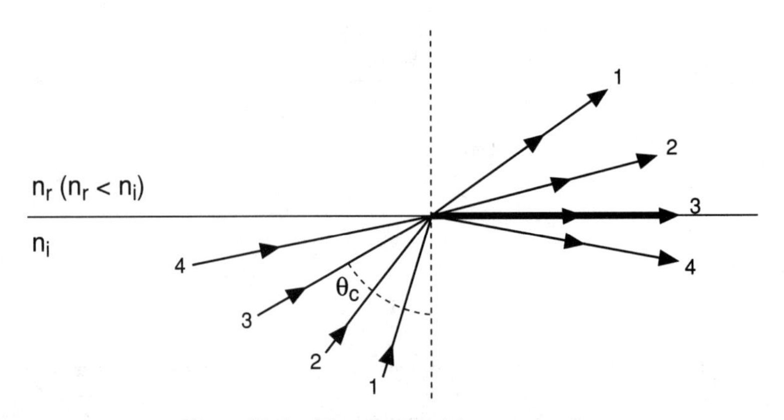

Figure 21-7. The concept of the critical angle.

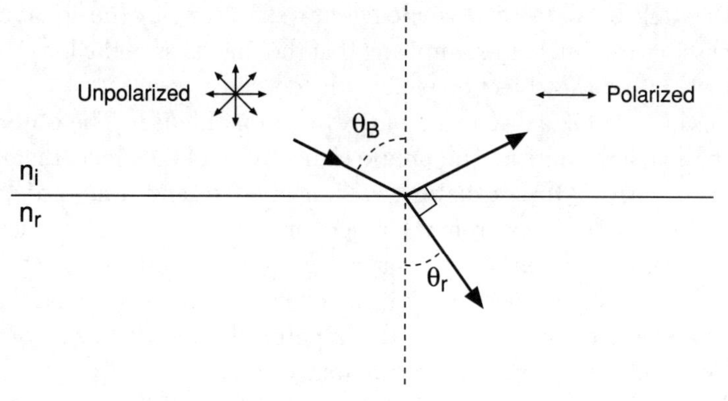

Figure 21-8. Production of polarized light by reflecting light off a surface with a greater index of refraction such that the reflected and refracted rays make a 90° angle.

Because $\sin \theta_r = 1$ when $\theta_i = \theta_c$, then

$$\sin \theta_c = n_r/n_i \text{ and } \theta_c = \arcsin (n_r/n_i)$$

Total internal reflection can only happen when the light rays are coming from within the higher index of refraction material. **It does not occur when rays pass from a low n to a high n.**

The idea of the critical angle is behind the use of fiber optics in medicine. In fiber optics, thousands of extremely thin glass fibers are fused into a rod. Light incident at one end of the rod is transmitted through the rod via total internal reflection and can be seen at the other end. Many medical procedures such as endoscopy, colonoscopy, and laparoscopy depend on the concept of total internal reflection.

In total internal reflection, the index of refraction outside the transmitting substance (e.g., fiber optic rods) is less than the index of refraction of the transmitting medium.

Recall that it is possible to produce polarized light by using polarized filters. **Another way in which to polarize light is to reflect light off of a surface with a greater index of refraction,** such that the reflected and refracted rays make an angle of 90° (Figure 21-8). Angle θ_B in Figure 21-8 is known as **Brewster's angle.** To determine Brewster's angle for a given interface between two media, use the formula:

$$\theta_B = \arctan (n_r/n_i)$$

Light incident on an interface will be partially reflected and partially refracted. If the incident light makes an angle equivalent to Brewster's angle with the normal vector to the surface, the reflected light will be totally polarized parallel to the interface between the two media.

V. Lenses and Mirrors

A. LENSES

CONVERGING (CONVEX) LENS

Lenses work by the property of refraction. When incident light rays encounter the lens, they are bent toward the normal vector. When the rays leave the lens, they are bent away from the normal vector. Because the normal vectors change across the lens and from one side of the lens to the other, the lens is able to **focus or diverge** light rays. To simplify the equations presented, all lenses in this discussion are thin lenses, implying that the thickness of the lens is less than the distance between the lens and the object on which it is focused.

Certain terms are used in a discussion of the function of lenses. The **object** is the item being examined with the lens or mirror. The **image** is the image of the object that is formed by light rays from the lens or mirror. **Object distances,** known as "p" and **image distances,** known as "q," are the perpendicular distances from the object or image to the lens or mirror.

Images can be **real or virtual.** A real image has light rays actually going through it, similar to the image formed by a slide projector. A virtual image, on the other hand, has no actual light rays travelling through it. It can be seen as the apparent location and size of the object when viewed via the lens or mirror creating the virtual image.

The **focal length** of a lens or mirror is the image distance if the object were held infinitely far away. In the ray diagrams that follow, solid arrows represent the object and dotted arrows represent the image. The **lens axis** is a line perpendicular to the surface of a lens and through its center.

To start ray tracing, choose the lens with which you have had the most experience—the convex (converging) lens, such as a magnifying glass. Just how it is possible to examine objects with this tool and have them appear bigger is illustrated in the ray tracing in Figure 21-9. The object is located at an object distance **p,** the image is located at the image distance **q,** and the focal points of the lens are marked with **f.** (The labels $\mathbf{d_o}$ for the object distance and $\mathbf{d_i}$ for the image distance are also common.)

In ray tracing, you draw light rays originating from the object and find out how the lens bends them. The intersection point of the traced rays identifies the position, orientation, and height of the image.

With the converging lens, you can draw three light rays. To start, it is important to realize that light rays from infinity are focused at f by a convex lens. Light rays from infinity are perpendicular to the plane of the lens, which makes them parallel to the lens axis.

The rays to draw:

1. A ray from the object that is parallel to the lens axis, refracted toward the focal point.
2. A ray travelling through the exact center of the lens. This ray will not encounter a net refraction as it passes through the lens, as the angle of the interface will not change after it passes through. To this ray, the lens is like a flat sheet of glass.
3. If needed, this ray is from the object passing through f on the same side of the lens as the object, refracted parallel to the lens axis.

Note that the intersection of these refracted rays gives you the location and height of the image. It is a real image because you have actual light rays that can come from the source to form

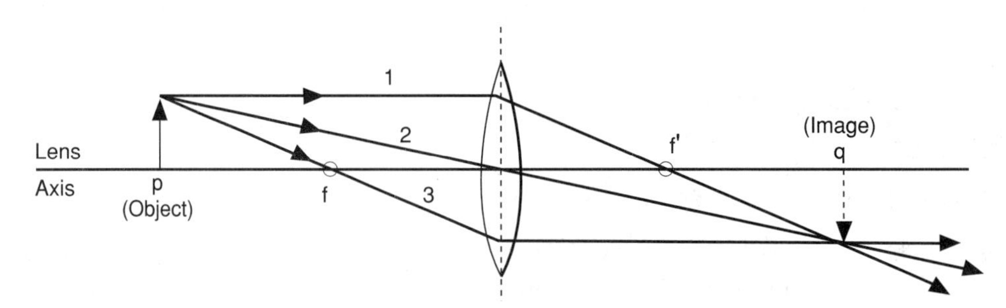

Figure 21-9. Ray diagram for a convex lens.

the image. Another important characteristic is that a **converging lens is defined as having a positive focal length.**

DIVERGING (CONCAVE) LENS

When tracing rays from a concave (diverging) lens (Figure 21-10), you have the same goal of tracing multiple rays from the object through the lens and seeing where they intersect. Again, the intersection point gives you the location and height of the image. The image of this ray tracing is different, however, because it is the image from which the light rays appear to be coming.

The rays to draw:

1. A ray goes straight through the center of the lens and is not refracted in any way. This ray can be treated in the same way as the unrefracted ray in the example of the converging lens.
2. A ray that is parallel to the lens axis hits the lens and diverges outward on a path that originates from the focal point f′.
3. A ray is aimed at the f point behind the lens and emerges from the other side of the lens as parallel.

The term f′ is the focal point of the diverging lens, and it is defined to be negative for a diverging lens. f′ is the point at which rays from infinity would seem to originate if an observer was looking from the other side of the lens.

If you draw rays 1 and 2 as just described and they do not intersect, think of this problem intuitively. The image is the point from which the diverged rays seem to be coming when viewed from the other side of the lens. So, draw the dotted line back from the refracted diverged ray back to f′. **The intersection of this dotted imaginary ray with the ray through the center gives you the location of the image.** In this diagram, as with that for the convex lens, the object distance is denoted p, and the image distance is denoted q. **The focal length f′ is a negative number.**

The image from a diverging lens is always Upright and Virtual. Virtual means it has no real light rays passing through it—it could not form an image on a screen. A good example of a virtual image is your reflection within a planar mirror; although it seems to be "behind" the mirror, no light waves are passing through that point.

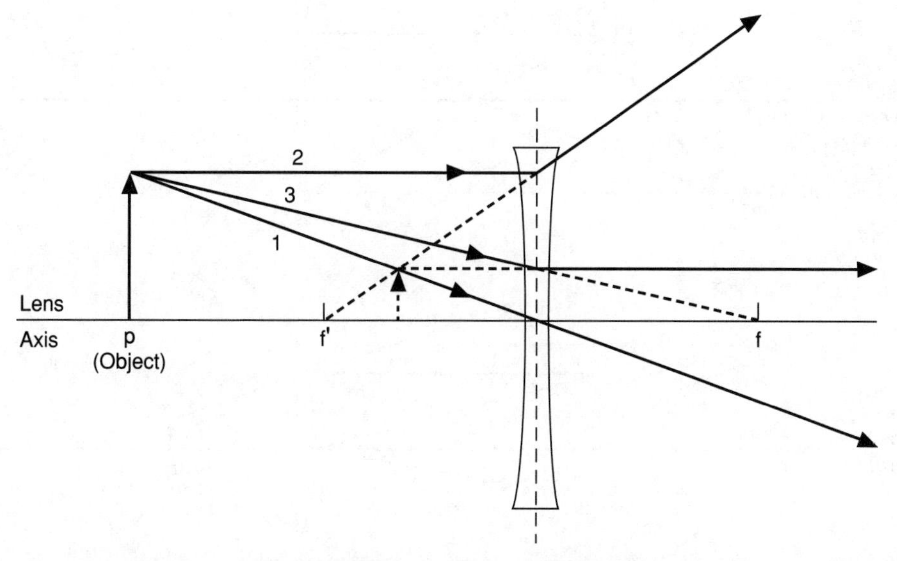

Figure 21-10. Ray diagram for a diverging lens.

B. MIRRORS

In lenses, rays are bent toward and away from the normal by **refraction,** allowing the ray to be bent when it passes through a medium of $n > 1$ with nonparallel surfaces. In mirrors, the light rays are bent during **reflection,** and the law of reflection determines their path. Planar, concave, and convex mirrors are shown in Figure 21-11.

PLANAR MIRROR

The planar mirror can be approached by ray tracing using the law of reflection.

The rays to draw:

1. A ray from the object that is parallel to the mirror axis (perpendicular to the plane of the mirror) and reflected directly backward.
2. A ray that hits the mirror at the lens axis and is reflected at an angle $\theta_r = \theta_i$, causing it to reflect downward.

Tracing these rays back using the dotted lines demonstrates where the rays appear to originate, which is the location of the image. Note that in a planar mirror, **the image is of equal height to the object and it is located the same distance away from the mirror ($q = p$).**

The **magnification** is the height of the image divided by the height of the object, and is **equal to one** for a planar mirror.

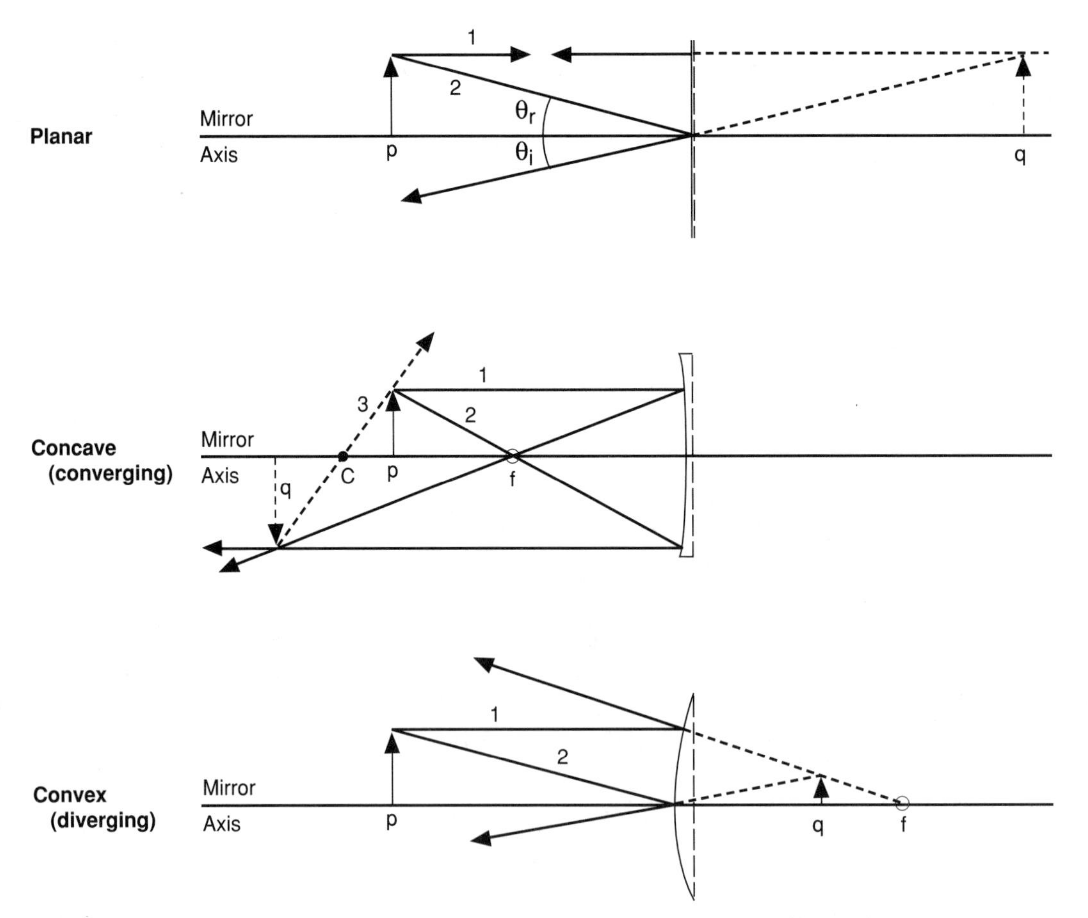

Figure 21-11. Ray diagrams for planar, concave (converging), and convex (diverging) mirrors.

CONCAVE (CONVERGING) MIRROR

A concave mirror acts to converge rays (see Figure 21-11, at center). An important new term for curved mirrors is the **radius of curvature (C).** By definition, C = 2f. A line from C to the mirror is always perpendicular to the mirror.

When a light ray hits the mirror, the incident angle, θ_i, and the reflected angle, θ_r, are both measured relative to the normal to the plane of the mirror; however, the normal to the plane of the mirror is also the line from **C** to the mirror. The focal point of a concave mirror is at 1/2 C, and it is the point to which a light ray parallel to the mirror axis is reflected.

The rays to draw:

1. A ray from the object parallel to the mirror axis, reflected through the focal point (1/2 C).
2. A ray from the object through the focal point, reflected back parallel to the mirror axis.
3. A ray from the object through the center of curvature. You may need to extend backward to find the image, especially when the object distance is greater than C.

Using these rays, it is possible to determine the location of the image (see Figure 21-11).

CONVEX (DIVERGING) MIRROR

A convex mirror acts to diverge rays. With a convex mirror, C and f are on the other side of the mirror from the object.

The rays to draw:

1. A light ray parallel to the axis of the mirror, reflected along a line originating at f. This ray can be traced back using the dotted line as shown.
2. A ray reflected across the mirror axis, which can also traced back. The intersection of the two traced back rays demonstrates the location of the image.

Similar to the diverging lens, the diverging (convex) mirror has a **D**iminished, **U**pright, **Vir**tual (DUV) image.

VI. Thin-Lens and Lens Maker's Equations

Ray tracing diagrams are a conceptual representation of what lenses and mirrors do. A simple equation describes the formation of images by mirrors and lenses. The **thin-lens equation** follows:

$$\frac{1}{p} + \frac{1}{q} = \frac{1}{f}$$

Remember that p is the object distance from the lens or mirror and q is the image distance from the lens or mirror. This valuable equation is used to find any object or image distance and any focal length if two of the three variables in the equation are known.

Another critical formula is the **magnification equation:**

$$m = \frac{-q}{p}$$

which can be used to find the magnification of an image **(m)** if q and p are known.

EXAMPLE 21-2

A CONVEX (CONVERGING) lens with a focal length of 10 cm is placed 0.5 m from an object. Find the location of the image. State whether the image is real or virtual.

SOLUTION:

Recall the thin-lens equation. Always remember that distances are in units of meters. Thus, $f = 0.1$ m and $p = 0.5$ m. Fill in the known values in the thin-lens equation: $1/0.5 + 1/q = 1/0.1$. Solving the equation gives $2 + 1/q = 10$; $1/q = 8$; $q = 0.125$ m. Note that q is positive, corresponding to a real image. A real image would be found on the opposite side of the lens as the object.

The thin lens equation is valid when the following rules are observed:

For lenses:

p + always
q + means the image is a real image, on the opposite side of the lens as the object
− means the image is a virtual image, on the same side of the lens as the object
f + for a converging lens (convex)
− for a diverging lens (concave)

For mirrors:

p + always
q + means the image is a real image in front of the mirror (on the same side as the object)
− means the image is a virtual image behind the mirror (on the opposite side as the object)
f + for a converging mirror (concave)
− for a diverging mirror (convex)

The following outline clarifies the relationships between the image and the object with different lenses and mirrors.

I. WITH A DIVERGING LENS (CONCAVE) OR MIRROR (CONVEX):

A. f is negative

B. The image is DUV (diminished, upright, and virtual), and q is negative (The image for convex mirrors is always diminished. For diverging lenses, the image is usually diminished).

II. WITH A CONVERGING LENS (CONVEX) OR MIRROR (CONCAVE)

A. f is positive

B. If $p > f$, the image is real, inverted, and q is positive

C. If $p < f$, the image is BUV (bigger, upright, and virtual), and q is negative

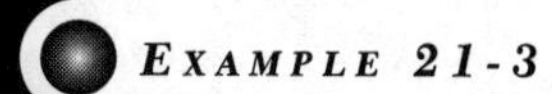

EXAMPLE 21-3

A FISH IN A fish tank swims by the bubbler, and takes a glance at his companion through one of the bubbles. Would the light rays from the companion be:

A. diverged as they pass through the bubble?
B. converged as they pass through the bubble?
C. unaffected as they pass through the bubble?

SOLUTION:
The light rays would be diverged as they pass through the bubble. The ray is bent away from the normal as it passes from the $n > 1$ water into the $n = 1$ air. Then, the ray is bent toward the normal as it passes from the $n = 1$ air into the $n > 1$ water again. Because the plane of the surface has changed, however, both of these effects act to diverge the light rays (Figure 21-12).

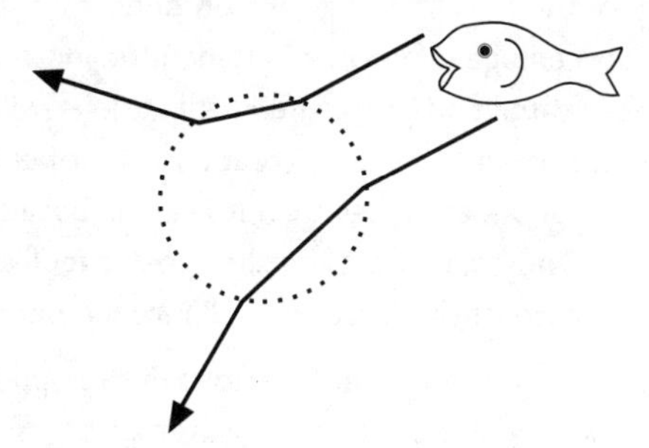

Figure 21-12. Concepts of light divergence through bubbles as discussed in Example 21-3.

To find the focal point of a lens, knowing just the index of refraction and the physical dimensions of the lens, use the **lens maker's equation:**

$$\frac{1}{f} = (n - 1)\left(\frac{1}{R_1} + \frac{1}{R_2}\right)$$

in which f is the focal length of the lens, n is the index of refraction of the material of the lens, and R_1 and R_2 are the radii of curvature of the two sides of the lens. Note that the definition of a radius of curvature for a concave side of a lens is negative, which allows a diverging concave lens to have a negative focal length. Also, it is possible to use this equation to find the focal length of a lens that has one convex and one concave side, such as lenses found in eyeglasses.

VII. Combinations of Lenses and Diopters

Combinations of lenses can be examined by using both equations and ray tracing. Ray tracing of combinations of lenses can be accomplished by using the image formed by one lens as the object of the other lens. For practice, try some of the combination of lenses problems in the review question section of this book.

EXAMPLE 21-4

YOU AND A FRIEND decide to invent a new way of seeing underwater. Instead of using a diving mask, which allows the cornea to converge light rays by providing an air interface, you devise different lenses to correct the diver's vision. You make a lens from glass with $n = 1.4$. Your friend shapes two thin glass plates, seals them, and then fills them with air.

1. Assuming that water has an index of refraction $n = 1.33$, whose design requires the greater radii of curvature?
2. What would be the shapes of the two lenses?

SOLUTION:

1. The ability of the lenses to bend light will depend on the ratio of their indices of refraction. From Snell's law, $n_1 \sin \theta_1 = n_2 \sin \theta_2$, the ratio of n_1 to n_2 describes the ability of the lens to bend light on entering and exiting the lens. Because of a greater percentage difference between the indices of refraction for air and water than for water and glass, the air-filled lens will bend light more effectively, and would not have to have as great a radius of curvature. Similarly, a glass lens is less powerful under water than it is in air because it is not able to bend light rays as well. The cornea acts as the most effective refractor of light rays (**not** the lens) because of the large ratio difference between n_{cornea} and n_{air}.
2. An air-filled lens under water functions in the opposite manner as a glass lens in air: a convex, air-filled lens under water diverges light rays and a concave, air-filled lens under water converges light rays. Thus, to counteract the diminished function of the cornea, you need a concave, air-filled lens.

Because its index of refraction is greater than that of water, the glass lens would function under water in the same manner as it would in air, albeit not as effectively. Thus, you would need a convex lens in order to converge the light rays.

An important concept of lenses in day to day life is **power. The power of a lens is simply the reciprocal of its focal length, 1/f.** Thus, the lens maker's equation gives the power of the lens on the lefthand side of the equation. Power is measured in **diopters** (equal to **1/meter,** because f is measured in meters). **The powers of lenses in combination are additive,** with the total power of the combined lenses being equal to the sum of the individual powers:

$$1/f_t = 1/f_1 + 1/f_2$$

This equation can be combined with the thin lens equation for each lens to allow the calculation of p or q for either lens, if given all other variables.

VIII. Dispersion

Dispersion is the property of a prism that allows it to scatter light of different frequencies. In a vacuum, light waves of all frequencies travel with the same speed. When they enter a medium

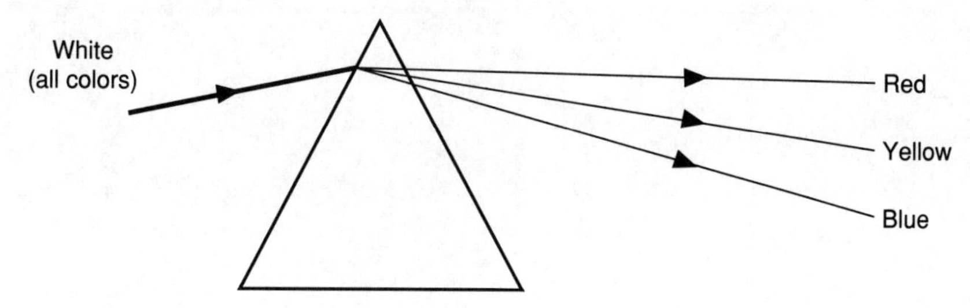

Figure 21-13. A prism separating light into its component colors (wavelengths).

with n > 1, however, light waves of different frequencies travel with slightly different speeds, varying by approximately 1 to 2% across the visual spectrum. The higher frequency waves tend to travel slower; therefore, the index of refraction of the prism for blue waves is slightly higher, and they will bend more on entering and exiting the prism. With the triangular cross section of the prism, separation of light of different frequencies results (Figure 21-13).

Dispersion is responsible for a prism separating light into its component colors (wavelengths) and for rainbows, in which water droplets are acting as the "prisms."

Atomic and Nuclear Structu...

The focus of this chapter is on the atom—its nucleus and the forces that bind it together. Other topics include radioactive decay and the photoelectric effect.

I. Basic Concepts

A. ATOMIC NUMBER AND ATOMIC WEIGHT

The physicist's basic view of the atom follows:

A = Atomic weight (4 in this example)
= Number of protons + number of neutrons

Z = Atomic number (2 in this example)
= Number of protons

$$^{4}_{2}\mathrm{He}$$

In this diagram, the atomic weight (A) is shown as the superscript over the element. It is the number of protons and neutrons in the atom. The atomic number (Z) is shown as the subscript under the element. It is the number of protons in the atom. **A − Z** is therefore equal to the number of neutrons in the atom.

B. NEUTRONS, PROTONS, AND ISOTOPES

Neutrons and protons are both nucleons, relatively large particles that make up the nucleus. Their weight is commonly measured in terms of atomic mass units, with **1 amu** being equal to 1/12 the weight of a ^{12}C atom (carbon with 6p, 6n, and 6e); 1 amu = 1.66×10^{-27} kg. A proton weighs 1.0073 amu, and the neutron is slightly heavier, at 1.0087 amu. Electrons are lighter, at 5.48×10^{-4} amu.

Neutrons are neutral; they have no electric charge. Protons have a positive charge of +e, 1.6×10^{-19} Coulombs. Electrons have a negative charge of −e. The typical radius of an atom, including its electron cloud, is about 10^{-10} m.

In **isotopes** of an element, the number of protons remains the same, but the number of neutrons differs. Thus, different isotopes of an element have the same atomic number, but different atomic weights.

II. Radioactive Decay

Radioactive decay is the process in which a nucleus emits particles and energy, decomposing into another particle or particles. The three types of radioactive decay are described.

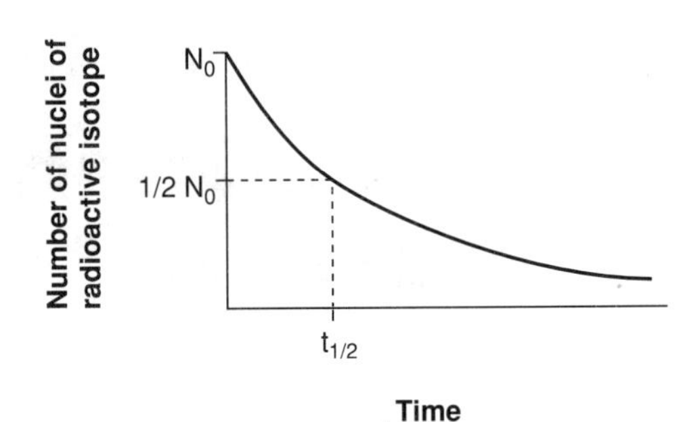

Figure 22-1. The initial number of nuclei of radioactive isotope (N_o) versus time. The half-life is the time required for ½ of the initial number of nuclei to decay.

1. In the first type, alpha (α) particles, which are helium nuclei, are released during radioactive decay. Their chemical symbol is:

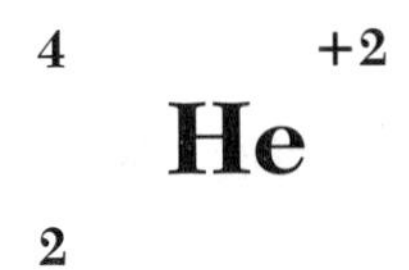

$$^{4}_{2}He^{+2}$$

They are relatively large and slow, and do not penetrate matter easily. Because paper shielding can stop alpha particles, they do not represent an external radiation hazard, but they can be hazardous if taken internally.

2. In the second type, beta (β) particles are released electrons ($-$e charge) or positrons (+e charge). They are stopped by shielding with approximately 1 cm of Lucite.
3. In the third type, gamma (γ) photons are released. These photons represent high energy, electromagnetic massless particles. Nuclei can emit gamma ray photons during radioactive decay. They penetrate matter easily, and require lead bricks as shielding.

Radioactive decay is examined in terms of the half life of a radioactive substance. **The half life ($t_{1/2}$)is the amount of time required for one-half of the nuclei in any sample of a given isotope to decay.** A graph of the number of radioactive nuclei with respect to time is presented in Figure 22-1. Note that the rate of radioactive decay is an exponential function.

The decay rate is the number of nuclei that decay per unit of time. N is the number of nuclei, and λ is the **radioactive decay constant,** which is related to the half-life. Thus, the change in the number of nuclei with respect to time is given by:

$$\frac{\Delta N}{\Delta t} = -\lambda N$$

A more important equation relates the decay constant and the half-life:

$$\lambda = 0.693 / t_{1/2}$$

III. Quantized Energy Levels for Electrons

This subject is addressed in the general chemistry review notes. As a brief overview, electrons circling a nucleus are confined to certain values (i.e., they are **quantized**). The energy level of an electron is determined by the principle quantum number, **n.** When an electron is taken away from the atom through ionization, the electron has been excited to the principle quantum number of n = infinity. Should the electron be excited by enough energy to change its principle quantum number, it will be held less tightly by the atom, but it will not be taken away.

IV. Mass Defect Principle and Nuclear Binding Energy

The mass defect principle states that the mass of every nucleus is less than the mass of the nucleons that make up that atom. Thus, the nucleus has a mass defect (Δm). The mass defect acts as a sort of a nuclear glue that gives stability to the atomic nucleus, preventing the protons' positive charge from pulling the nucleus apart.

The mass defect of carbon-12 is found from the atomic composition as follows: 6 protons (6×1.0073 amu) + 6 electrons (6×0.005 amu) + 6 neutrons (6×1.0087 amu) gives a total mass of 12.099 amu. Because carbon-12 has an atomic mass of 12.000 amu, there is a mass defect (Δm) of 0.099 amu.

The nuclear binding energy is the energy equivalent of the mass defect. It is found by using Einstein's formula:

$$E = (\Delta m)c^2$$

in which E is measured in Joules, m in kilograms, and c, the speed of light, in meters per second. The energy of 1 amu according to this formula is: (Note 1 eV = 1.6×10^{-19} J).

$$\begin{aligned} E &= 1 \text{ amu} \times c^2 \\ &= (1.66 \times 10^{-27} \text{ kg})(3 \times 10^8 \text{ m/sec}) \\ &= 1.49 \times 10^{-10} \text{ J} \\ &= 1.49 \times 10^{-10} \text{ J}/1.6 \times 10^{-19} \text{ J} = 9.3 \times 10^8 \text{ eV} \end{aligned}$$

The binding energy per nucleon gives a good idea of the stability of the nucleus. It shows how much "glue" is spread across each nucleon.

V. Photoelectric Effect and Fluorescence

A. PHOTOELECTRIC EFFECT

A photon is a massless bundle of electromagnetic energy. The energy of a single photon can be described by the formula:

$$E = hf$$

in which E is the energy, h is a constant called Planck's constant, and f is the frequency. Thus, the energy of photons varies with frequency.

The photoelectric effect occurs when electrons eject from the surface of a metal when it is irradiated with electromagnetic radiation. For this effect, light of one frequency and wavelength (**monochromatic** light) strikes a metal plate. If an electron in the metal absorbs enough energy from the photons of light, it can escape the metal's surface. The electrons moving from the plate can be detected by a collector, thus illustrating a photoelectric current. The kinetic energy of the emitted electrons would be equal to:

$$1/2mv^2 = hf - \phi$$

in which **ϕ**, the work function, is the minimum amount of energy needed to escape the surface of the metal. All of the rest of the energy obtained from the photon is translated into the kinetic energy of the photon.

B. FLUORESCENCE

Fluorescence is the property that allows electrons to emit light as they lose energy falling from an excited state to a less excited state or the unexcited ground state.

HIGH-YIELD REVIEW QUESTIONS

Section I: Physics

Vectors, Kinematics, Statics, and Force

In the following questions, unless stated otherwise, neglect air resistance.

1. The SI system of fundamental units includes:

A. the foot, pound, and second.
B. the meter, kilogram, and second.
C. the meter, liter, and second.
D. the foot, quart, and second.

2. What is the difference between vector A and vector B?

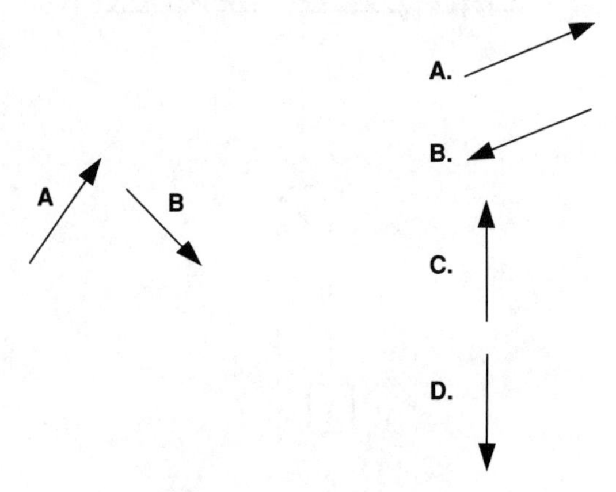

3. A ball rotaing in a circular path at the end of a string is moving at a constant speed. The ball would be described as:

A. accelerating.
B. having constant velocity.
C. both of the above.
D. neither of the above.

4. Which ones are scalars?

I. Distance
II. Velocity
III. Speed
IV. Displacement
V. Acceleration

A. I and III
B. I, III, and V
C. III and V
D. I and V

5. A ball is thrown vertically upward, reaches its highest point, and falls back down. Which statement is true?

A. The acceleration is always in the direction of motion.
B. The acceleration is always opposite the direction of motion.
C. The acceleration is always up.
D. The acceleration is always down.

6. A ball is thrown upward. During its flight, its acceleration:

A. increases.
B. decreases.
C. stays the same.
D. is zero.

7. Two objects, A and B, travel in the paths shown. Object A travels with a uniformly increasing speed, whereas object B travels with a uniformly decreasing speed. Which arrows show the acceleration of the objects?

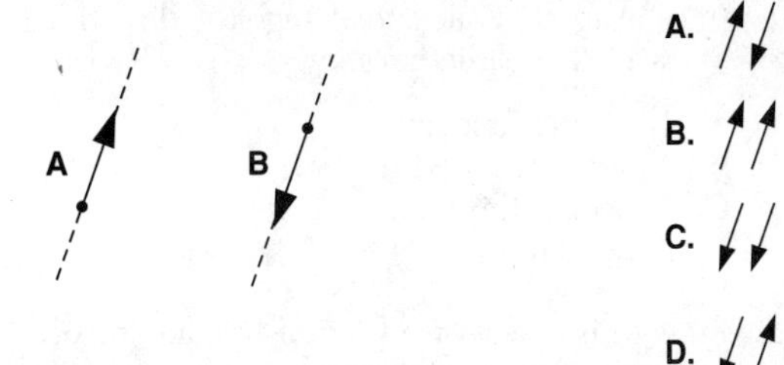

. An object is shot from a cannon in the path shown. Which statement is true?

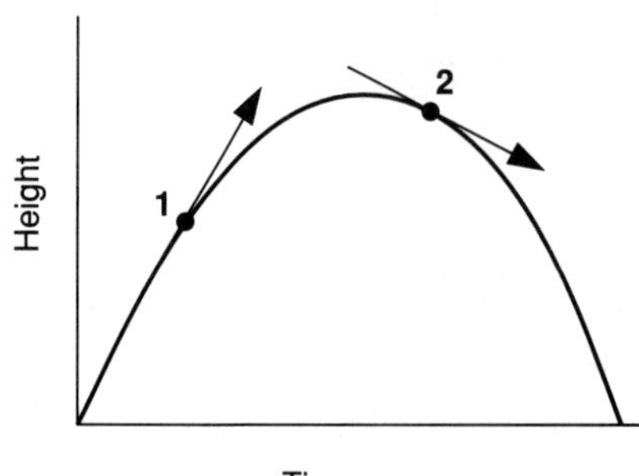

A. The object has different accelerations at points 1 and 2.
B. The object has the same acceleration at points 1 and 2.
C. No comparison can be made.
D. The object has greater acceleration at point 1.

9. An object falls 50 meters from rest. What information is needed to calculate its final velocity?

A. Initial velocity
B. The acceleration of the object
C. The weight of the object
D. No additional information

10. A sled travels 5 m/s on a friction-free air track. A force, F, acts to change the direction of travel. This situation shows:

A. Newton's first law.
B. Newton's second law.
C. Newton's third law.
D. none of these.

11. Mass M is at rest on a friction-free incline of 30 degrees. Mass M is connected by a string to a mass of 2 kg as shown. Find the mass of M. Assume $g = 10\ m/s^2$.

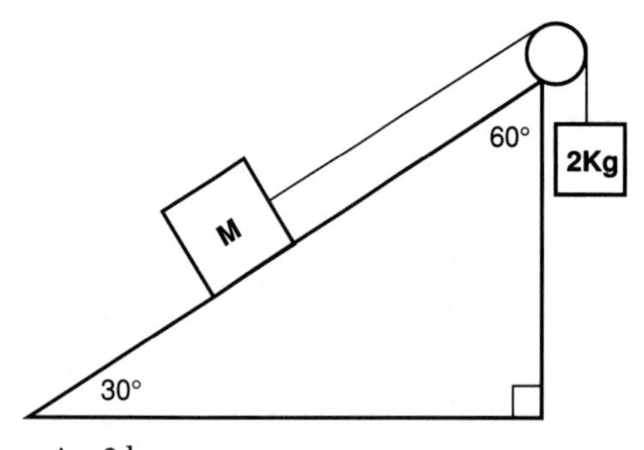

A. 2 kg
B. 4 kg
C. 6 kg
D. 8 kg

12. Two masses of 2 and 3 kg, respectively, are accelerated at $20\ m/s^2$ uniformly on a frictionless surface. The difference between tension T_2 and T_1 is:

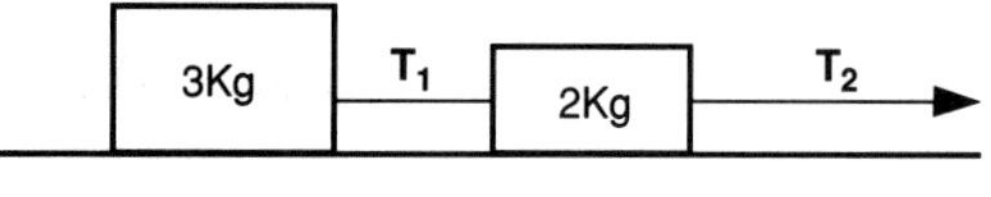

A. 20 N.
B. 10 N.
C. 60 N.
D. 40 N.

13. Masses X and Y rest on the massless and frictionless pulley as shown. The mass of X is greater than the mass of Y. What is the acceleration of the system?

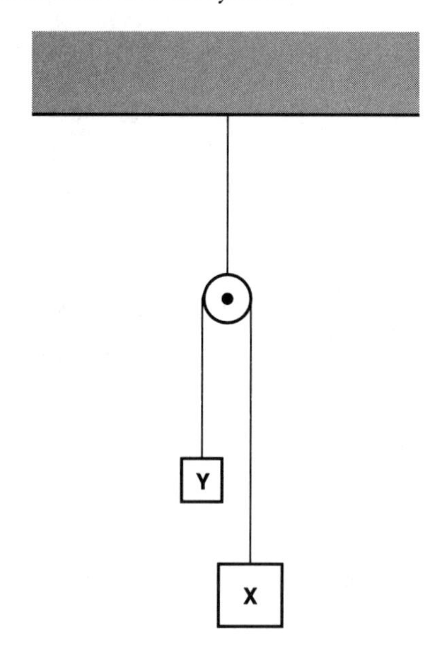

A. g
B. $((X - Y)/XY)(g)$
C. $((X - Y)/(X + Y))(g)$
D. $((X + Y)/(X - Y))(g)$

14. Calculate the unknown tension T.

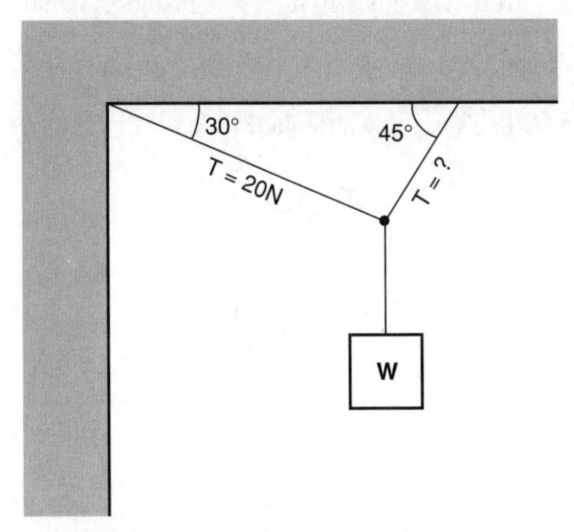

A. $10\sqrt{3}$
B. $10\sqrt{6}$
C. $5\sqrt{3}$
D. $5\sqrt{6}$

15. Two blocks, of masses m_1 and m_2, hang from a rope of negligible mass. The tension in rope X is:

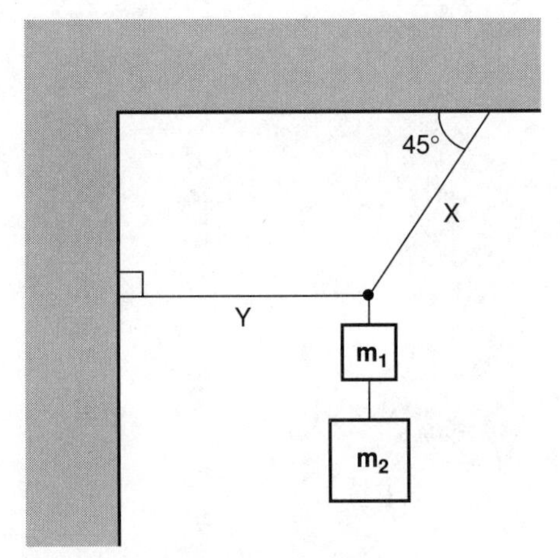

A. $(m_1 + m_2)(g)(\sin 45°)$.
B. $\{(g)(\sin 45°)\}/(m_1 + m_2)$.
C. $\{(m_1 + m_2)(g)\}/(\sin 45°)$.
D. unable to be determined without further information.

16. A 50-N rod with a center of mass marked "C" is placed on a fulcrum as shown. What happens?

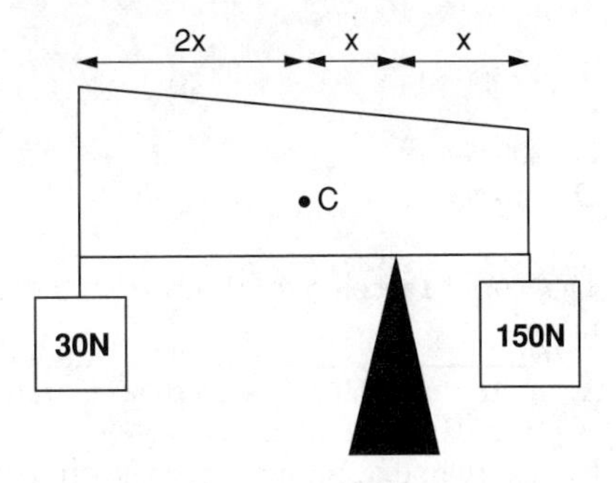

A. The system remains in equilibrium.
B. The system tilts to the right.
C. The system tilts to the left.
D. The answer cannot be determined from the information given.

17. Which factor(s) is (are) NOT vector quantities?

I. Power
II. Velocity
III. Work
IV. Displacement
V. Acceleration

A. I, III, and IV
B. III
C. III and IV
D. I and III

18. Where mass = m, length = l, and time = t, power is equivalent to:

A. ml^2/t.
B. m^2l/t^2.
C. ml^2/t^3.
D. ml^2/t^2.

19. When a stone is thrown in an upward trajectory:

I. it gains PE as it rises.
II. it loses PE as it rises.
III. it gains KE as it rises.
IV. it losses KE as it rises.
V. work is accomplished against gravity in throwing the stone.

A. I and IV
B. II and IV
C. II, IV, and V
D. I, IV, and V

20. A crane lifts a 200-kg block upward a distance of 10 meters. The rate of work is a steady 200 W. How long does it take to lift the block? Assume $g = 10 \text{ m/s}^2$.

A. 50 sec
B. 100 sec
C. 200 sec
D. 250 sec

21. The work to accelerate a body from 0 to 10 m/s is:

A. more than that required to accelerate it from 10 m/s to 20 m/s.
B. less than that required to accelerate it from 10 m/s to 20 m/s.
C. equal to that required to accelerate it from 10 m/s to 20 m/s.
D. any of the above depending on the time needed to change the speed by accelerating.

22. Two people lift identical blocks of mass X. Person A lifts the block vertically to a height H. Person B slides the block up an incline plane with frictionless rollers to a height H. Who does more work?

A. Person A
B. Person B
C. Both do the same work.
D. Neither does any work.

23. Two blocks, A and B, rest on frictionless inclines with base lengths, angles of inclination, and masses given. The blocks are now released from rest. Which equation about the speeds of A and B at point R is true?

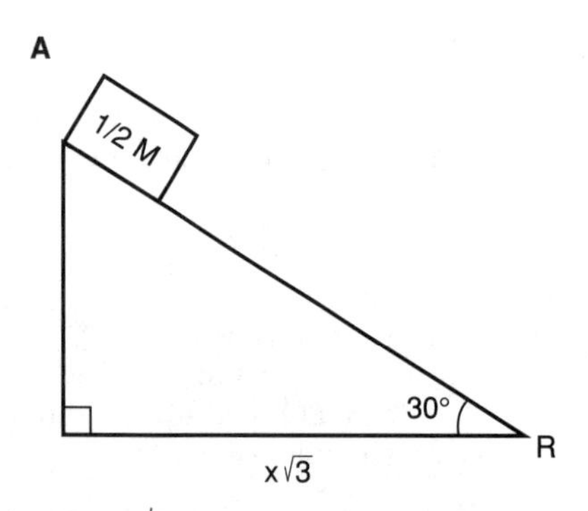

A. $V_a = \sqrt{2}V_b$
B. $V_b = \sqrt{2}V_a$
C. $V_a = \sqrt{3}V_b$
D. $V_b = \sqrt{3}V_a$

24. A 5-kg ball rolls down a 45-degree hill from rest. The hill is 10 meters high and 14 meters long. The coefficient of friction for the ball is 0.2. Assume $g = 10 \text{ m/s}^2$. How much energy, in joules, is dissipated as heat during the ball's trip down the hill?

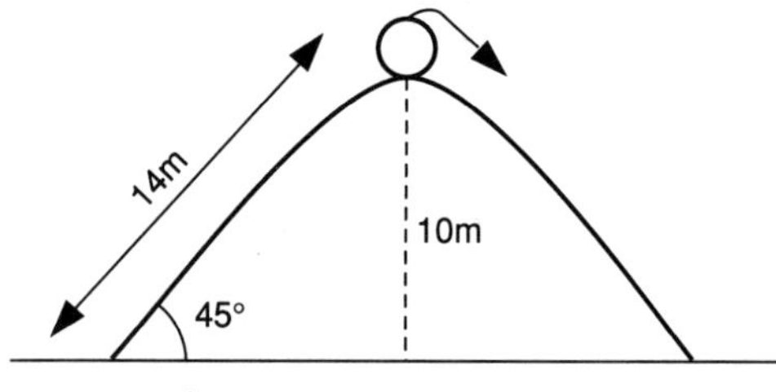

A. $50\sqrt{2}$
B. $70\sqrt{2}$
C. $200\sqrt{2}$
D. $300\sqrt{2}$

25. A sled of mass M is at rest on a ramp X meters high. It loses R joules of heat as friction when it slides down the ramp. What is its KE in joules at the bottom of the ramp?

A. $R - MgX$
B. MgX
C. $MgX - R$
D. None of the above

SOLUTIONS

Vectors, Kinematics, Statics, and Force

1. **B** Choice A is incorrect because it gives measurements in the British system of units. Choices C and D give measurements that are not included as fundamental units of either the SI or the British system. Units of volume are not considered fundamental units.

2. **C** The difference between two vectors is obtained by adding the first vector (A) to the negative of the second (B). The negative of a vector has the same magnitude, but opposite direction. To answer this question, connect the tail of B with the head of A and draw the resultant vector by connecting the tail of A with the head of −B:

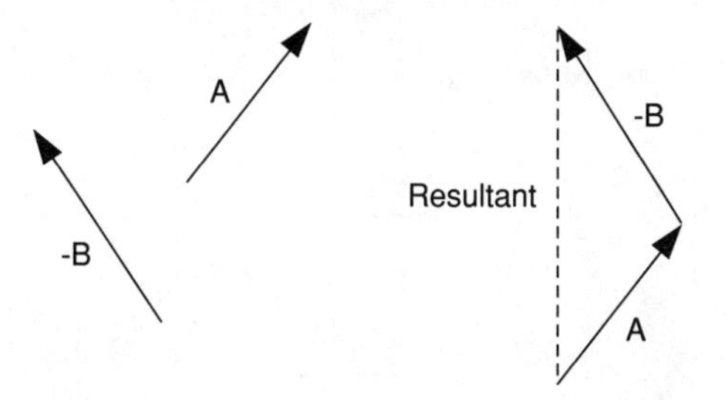

3. **A** The ball has constant speed, but its direction of travel is continually changing as it rotates. Because the direction constantly changes, so does the velocity; the ball is therefore accelerating. Recall that acceleration = $\Delta v/\Delta t$.

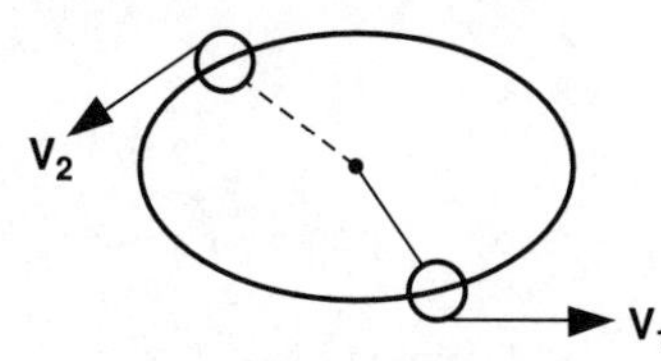

4. **A** Distance and speed are scalars. The other choices, including displacement, are all vector quantities.

5. **D** The only force acting to accelerate the ball is gravity, which always acts to accelerate the ball toward the earth.

6. **C** The ball is accelerated while it is in the hand. However, once it leaves the hand, the only force accelerating the ball is gravity. Gravity acts as a constant, downward force.

7. **B** Uniformly increasing speed accelerates in the direction of travel. Uniformly decreasing speed decelerates in the direction of travel. However, a deceleration can be interpreted as an acceleration in the opposite direction because a resistance force is necessary to decrease the speed of the object. This resistance force is related to acceleration by F = ma.

8. **B** The acceleration of gravity (g) is the only force acting on the object once it leaves the cannon. In these problems, ignore air resistance.

9. **D** Use the equation: $v_f^2 - v_o^2 = 2\ ax$. The final velocity can be calculated by knowing the initial velocity, acceleration, and distance traveled. Since v_o is zero in this question, $v_f = (2ax)^{1/2}$. The only other information needed to solve the problem is the distance traveled (which is given as 50 m) and g.

10. **A** Newton's first law states that a body continues in a state of rest, or uniform motion is a straight line, unless it is compelled to change that state by outside forces. This first law is being described in this problem. Newton's second law says that F = ma. Newton's third law states that for every action there is an equal and opposite reaction.

11. B The diagram shows the net forces acting on the blocks. Because the block system is at rest, the vertical forces acting on blocks 1 and 2 are equal.

$mg \sin 30° = mg$

(block 1) (block 2)

$m(10)(1.2) = (2)(10)$

$5\ m = 20$

$m = 4$ kg

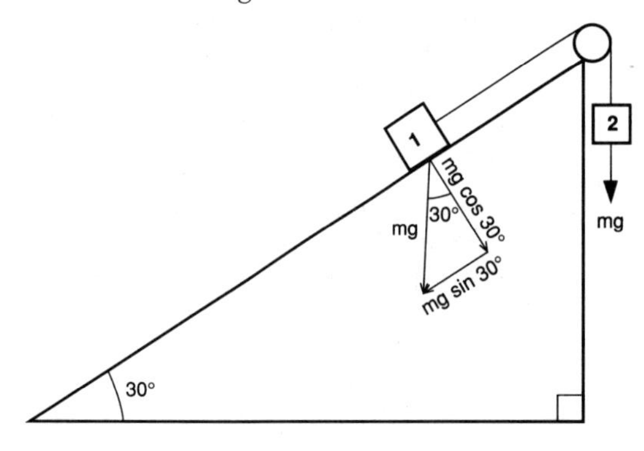

12. D Solve the problem using the following equations:

$T_1 = ma = (3\text{ kg})(20\text{ m/s}^2) = 60\text{ kgm/s}^2$

$T_2 = (3\text{ kg} + 2\text{ kg})(20\text{ m/s}^2) = 100\text{ kgm/s}^2$

$T_2 - T_1 = 40\text{ kgm/s}^2$ or 40 N.

13. C $T_1 = T_2$, because the tension in a rope is equivalent on both sides of a pulley. In the next step, write $F = ma$ for each block. Based on the relative magnitude of the masses, the prediction would be that $Xg > T_2$ and $T_1 > Yg$.

$F = ma$

$T_1 - Yg = Ya$

$Xg - T_2 = Xa$

Because $T_1 = T_2$, add the equations and cancel T_1 and T_2.

$Xg - Yg = Ya - Xa$

$(X - Y)g = (X + Y)a$

$a = \{(X - Y)/(X + Y)\}(g)$

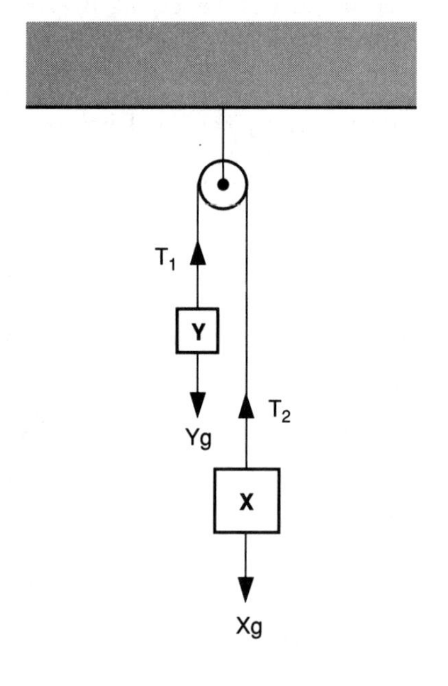

14. **B** The sum of the x-components must be zero. First, draw a diagram showing all the forces. Then, set up an equation in which all the x-components of force add to zero. Solve the equation for the unknown T.

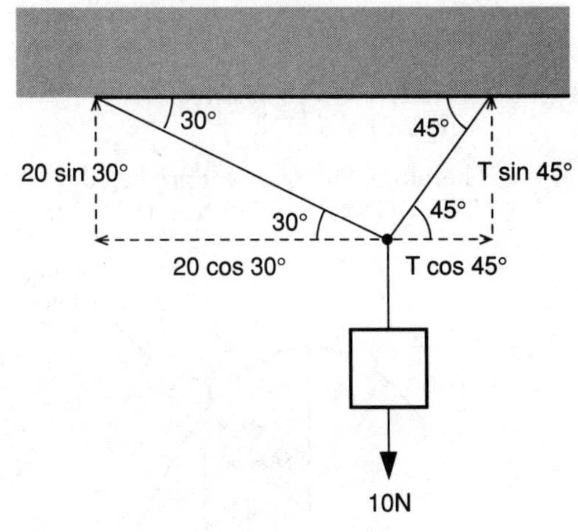

$T \cos 45° - 20 \cos 30° = 0$

$(\sqrt{2}/2)(T) = (20)(\sqrt{3}/2)$

$T = 10\sqrt{6}$

15. **C** Because the system is in equilibrium, the y-component of tension in rope X must equal the weights of masses m_1 and m_2.

$X \sin 45° = (m_1 + m_2)g$

$X = \{(m_1 + m_2)(g)\}/\sin 45°$

16. **B** The forces on the left of the fulcrum must equal the forces on the right to balance. More specifically, the downward forces to the left of the fulcrum produce a counterclockwise torque. The downward forces to the right of the fulcrum produce a clockwise torque.

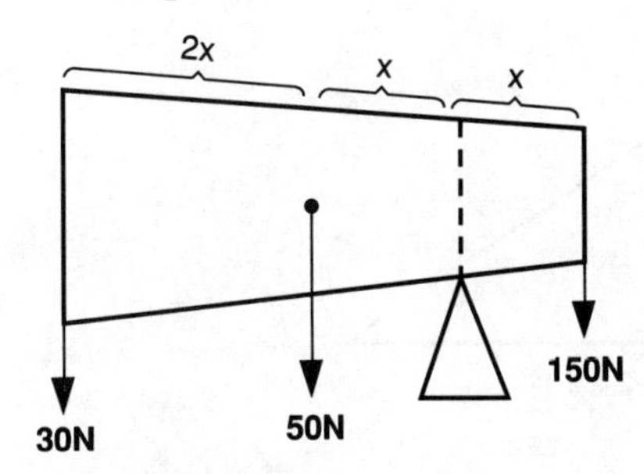

Multiply force by distance from the fulcrum.

$(3x)(30\ N) + (x)(50\ N) = (x)(150\ N)$

$(x)(140\ N) = (x)(150\ N)$

Because the (force–distance) product is greater on the right, the system tilts to the right.

17. **D** Work is the product of force and distance, where the force is directed in the same direction as travel.

$W = Fs \cos \theta$, where F = force, s = distance, θ = angle between force and direction of travel.

Work is a scalar, given in joules. 1 J = 1 Nm. Power is the rate at which work is done, or $P = \Delta W/\Delta t$.

18. **C** Consider just units here. Power = W/t = Fs/t.

Because F = ma, and distance(s) = 1, power = mal/t.

Because $a = (1/t^2)$, power = $\{m(1/t^2)(1)\}/t = ml^2/t^3$.

19. **D** A stone thrown upward gains PE as it rises because PE = mgh. Let h be a height above a reference point (the earth). This increase in PE corresponds to work done against gravitational attraction. Furthermore, as PE increases, KE correspondingly decreases. Having reached its peak, the stone gains speed and KE as it falls and its PE decreases.

20. **B** The work to lift a block is equivalent to the increase in PE. Therefore, W = mgh or $(200\ kg)(10\ m/s^2)(10\ m) = 20{,}000$ Nm or 20,000 J. The rate of work is power, and is given as 200 W. Because power = $\Delta W/\Delta t$, 200 W = 20,000 J/t. Solving this equation for t gives the time required to lift the block: t = 100 seconds.

21. **B** Two solutions to this problem are given here.

Solution 1: Work = (F)(s) = (m)(a)(s). Therefore, work depends on the mass of the object, its acceleration, and the distance through which the object accelerates. Assume that the net accelerations ($10\ m/s^2$) are equivalent in all the choices, as is the mass of the body. However, objects that travel at greater speeds (as in choice A) travel greater distances over a given period of time. Distance traveled relates to work: W = (m)(a)(s). This equation shows why it requires more work to accelerate at high speeds.

Solution 2: Another way to solve this problem is to use the work–energy theorem. It states that W = ΔKE + Δ PE. Here, because PE is not mentioned, one can assume that it remains constant.

For 0 –> 10 m/s, $W = \Delta KE = ½\, m(10^2 - 0) = ½\, m(100)$.

For 10 –> 20 m/s, $W = \Delta KE = ½\, m(20^2 - 10^2) = ½\, m(300)$.

Therefore, choice B is correct.

22. **C** This question is difficult because it requires a thorough understanding of the concepts. Person B pushes the block up a frictionless incline such that all force applied is converted to the vertical displacement of the block up the incline. Do not be confused by the fact that person B must push the block a greater distance than person A lifts it. Remember that the only force that is overcome by sliding up this frictionless incline is gravity, the same as in lifting the block. Because work is defined as a force applied over a distance and the only applied force is to overcome the vertical force of gravity, consider only the net vertical distance covered by the block. Therefore, lifting the block requires the same work as sliding it up a frictionless incline.

23. **A** Use conservation of energy to answer this question. The PE of the blocks at rest must equal the KE at the bottom of the ramps.

KE = PE

$½\, mv^2 = mgh$

$v = (2\, gh)^{1/2}$

Now, find h for both ramps. Use the 30–60–90 rule.

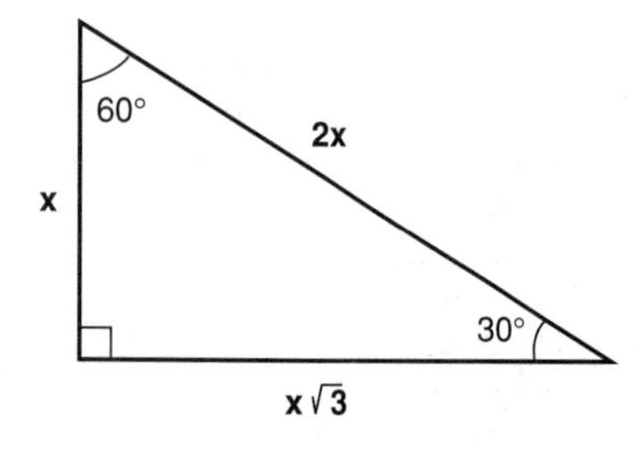

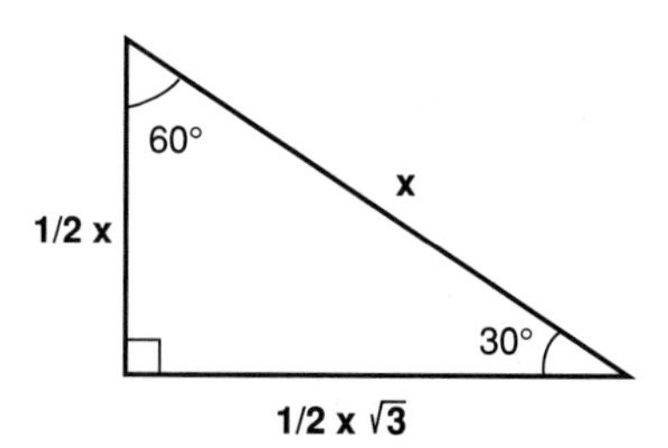

$v_a = (2\, gh)^{1/2} = (2\, gx)^{1/2}$

$v_b = (2\, gh)^{1/2} = [(2\, g)(x/2)]^{1/2} = (gx)^{1/2}$

Therefore, $v_a = v_b\sqrt{2}$.

24. **B** The components of force are given in the diagram. Because heat production is due to friction produced over a distance, the friction force acting on the ball is:

$f = \mu mg \cos\theta$, where $f = \mu N$.

Substituting values from the problem gives:

$f = (0.2)(5\text{ kg})(10\text{ m/s}^2)(\cos 45°) = (10)(\sqrt{2}/2) = 5\sqrt{2}\text{ N}$.

Therefore, the total heat produced = $(5\sqrt{2}\text{ N})(14\text{ m}) = 70\sqrt{2}\text{ Nm}$ or $70\sqrt{2}$ J.

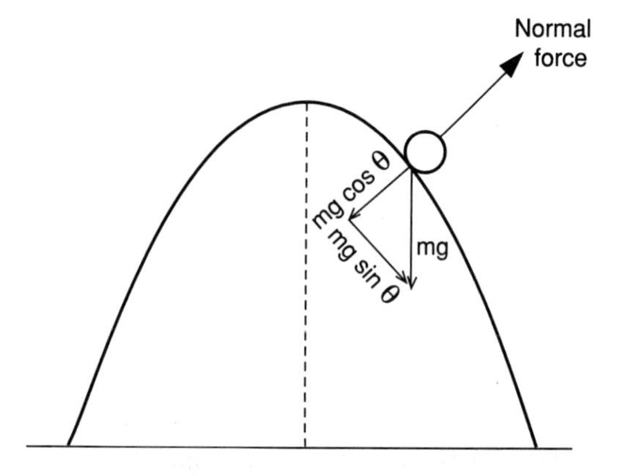

25. **C** At the top of the ramp, the sled has all PE; therefore, PE = mgh = MgX. This PE is converted to heat and KE by the time the sled reaches the bottom of the ramp. PE = heat + KE. Therefore, KE = PE − heat, and KE = (MgX − R) joules.

HIGH-YIELD REVIEW QUESTIONS

Section I: Physics

Force, Momentum, Circular Motion, and Gravitation

In the following questions, unless stated otherwise, neglect air resistance.

1. A 1-kg object travels with constant velocity at 2 m/s in direction X. An unknown force is applied to the object along direction X. The object speeds up, and now travels 35 meters in 5 seconds. What is the unknown force?

A. 4 N
B. 1 N
C. 2 N
D. 1.5 N

2. A 10-gram mass is dropped to the ground from 2 meters. A 5-gram mass is dropped from 3 meters. What is the difference in final velocities of the masses when they hit the ground?

A. $\sqrt{6g} - 2\sqrt{g}$
B. $\sqrt{6g} - 2$
C. $2\sqrt{g} - 6\sqrt{g}$
D. $\sqrt{2g} - 6\sqrt{g}$

3. Calculate T_1 and T_2.

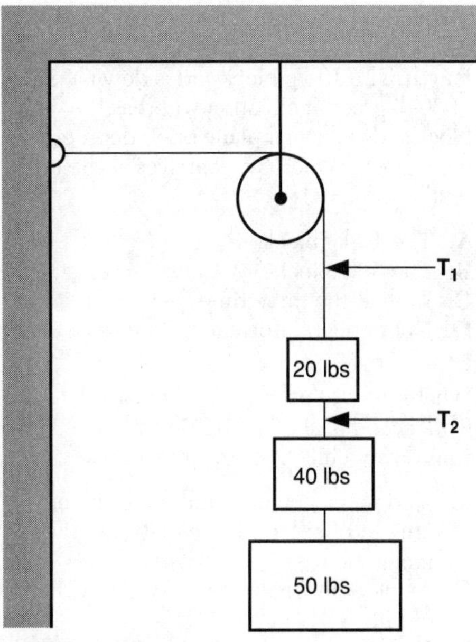

A. 110 lb, 20 lb
B. 55 lb, 20 lb
C. 110 lb, 90 lb
D. 55 lb, 90 lb

4. A man stands on a scale in an elevator that is dropped in free fall. His mass is 200 kg. How much does the scale read? Assume $g = 10$ m/s^2.

A. 2000 N
B. 200 N
C. 20 N
D. 0 N

5. A boy hangs from a spring scale attached to the ceiling of an elevator. The spring scale shows that he weighs 64 lb. When the elevator accelerates upward at 4 ft/s^2, how many pounds does the boy weigh? Use $g =$ 32 ft/s^2.

A. 64 lb
B. 60 lb
C. 68 lb
D. 72 lb

6. An elevator accelerates downward at 5 m/s^2. A woman standing in this elevator weighs 50 N. What is the woman's weight when the elevator is at rest?

A. 25 N
B. 100 N
C. 45 N
D. 75 N

7. Two blocks of masses 10 kg and 5 kg slide down frictionless incline planes of the same length. The 10-kg block slides down a plane of 30 degrees' inclination whereas the 5-kg block slides down a plane of 60 degrees' inclination. Which block arrives at the bottom first?

A. The 10-kg block
B. The 5-kg block
C. Both at the same time
D. Not enough information to determine

8. What information is needed to calculate the work accomplished by friction when a sled runs down a hill? Assume g = 10 m/s^2.

A. Sled mass, height of hill, final velocity
B. Height of hill, initial velocity, angle of inclination
C. Angle of inclination, final velocity, height of hill
D. Final velocity, sled mass, initial velocity

9. Impulse, as it refers to the force acting on an object, is:

I. mv.
II. F(Δt).
III. a vector.
IV. a scalar.

A. I and III
B. II and IV
C. I and IV
D. II and III

10. Which force(s) is(are) conserved in inelastic collisions?

I. Momentum
II. PE
III. KE
IV. Velocity

A. II and IV
B. I and III
C. I
D. III

11. An object of mass 2 kg travels with a speed of 4 m/s in the $^+$x direction on a frictionless surface. It collides with a 3-kg object that was at rest. The collision is inelastic. Following the collision:

A. $v_f = 7/6$ m/s.
B. $v_f = 8/5$ m/s.
C. $v_f = 5/6$ m/s.
D. $v_f = 4/5$ m/s.

12. Following an inelastic collision, two blocks of masses 5 kg and 9 kg travel in the $^+$y direction with a speed of 10 m/s. The KE of the system is:

A. 300 J.
B. 500 J.
C. 600 J
D. 700 J.

13. A block initially at rest explodes into three parts of equal mass M. Part 1 and part 2 have velocity vectors perpendicular to each other with equal speed of X. What is the speed of part 3?

A. $x\sqrt{2}$
B. $x\sqrt{3}$
C. $\sqrt{2}/x$
D. $\sqrt{3}/x$

14. Which one is NOT true of elastic collisions?

A. They conserve momentum.
B. They conserve KE.
C. They obey Newton's third law.
D. None of the above; all three are true.

15. A bullet is fired into a block of wood hanging from a string as shown in the following diagram. The bullet has mass x, and the block has mass y. The bullet initially has a speed v before impact. After bullet impact, the bullet–block system swings h meters vertically. What is the KE of the bullet–block system at impact?

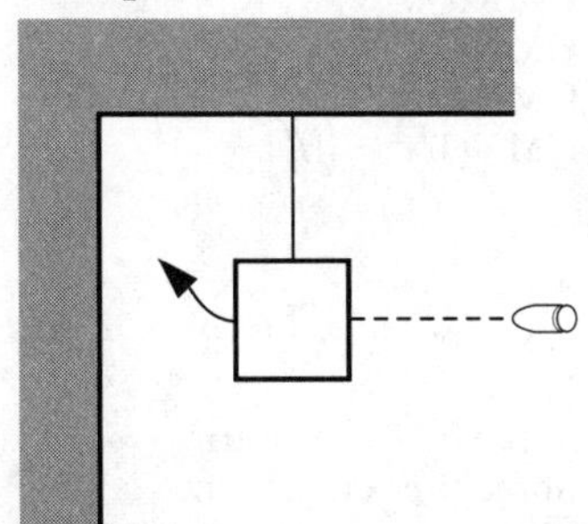

A. ½(x + y)v
B. (x + y)gh
C. ½(x + y)h
D. ½vg(x + y)

16. An arrow is shot into a block at rest on a frictionless table. The arrow has mass 0.1 kg, whereas the block has a mass of 0.9 kg. The arrow hits the block with a velocity of 100 m/s. What is the speed of the block after being penetrated by the arrow?

A. 1 m/s
B. 10 m/s
C. 15 m/s
D. 20 m/s

17. A 5-kg block travels at 10 m/s along a frictionless surface. It strikes a resting 2-kg block in an elastic collision. What is the situation after the collision?

A. Total KE and momentum are the same as before the impact.
B. Total KE and momentum are different than before the impact.
C. Momentum is shared equally between the two blocks.
D. Momentum and PE are both conserved.

18. An object rotates at constant speed in a circular path of radius r. The object can be said to have:

I. tangential velocity.
II. tangential acceleration.
III. centripetal acceleration.
IV. angular velocity.

A. I and II
B. III and IV
C. I, III, and IV
D. I, II, III, and IV

19. An object rotates in a circular orbit about a point with a constant speed. Which acceleration vector depicts its net acceleration?

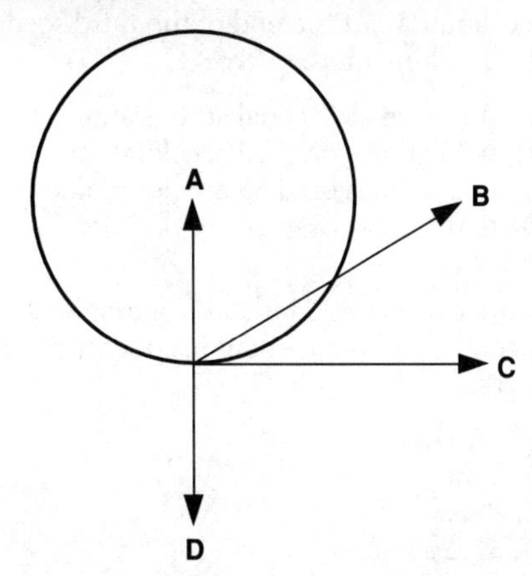

A. A
B. B
C. C
D. D

20. A record rotates 10π radians in 5 seconds. Its angular velocity is:

A. ½π radians/sec.
B. 2π radians/sec.
C. 5π radians/sec.
D. ¼π radians/sec.

21. A point on the outside of a rotating tire travels 10π meters. The radius of the tire is ½ meter. What angle does the point rotate through?

A. 2700 degrees
B. 1800 degrees
C. 3600 degrees
D. 4900 degrees

22. A point on a rotating flywheel has a higher tangential acceleration if:

A. it is closer to the center of the flywheel.
B. it is farther from the center of the flywheel.
C. it has greater mass.
D. acceleration of the point is independent of its location.

23. A rotating tire on a car decelerates at a constant rate owing to application of the car's brakes. After the first second, its angular acceleration has fallen to ½ its original acceleration. At the end of the third second, the acceleration is:

A. 0.125 times original acceleration.
B. 0.250 times original acceleration.
C. 0.225 times original acceleration.
D. 0.325 times original acceleration.

24. A tire completes 5 radians' rotation in 1 second. How many rotations does it complete in 5 seconds?

A. 10π
B. 25π
C. $10/\pi$
D. $25/2\pi$

25. Planet X has a mass ½ and a radius ¼, respectively, of those of the Earth. What fraction of the Earth's acceleration due to gravity (g_e) is g_x?

A. ¼ of that on Earth
B. ⅛ of that on Earth
C. 8 times that on Earth
D. 4 times that on Earth

26. Planet A has mass M, wheres planet B has mass N. They are X units apart and moving away from each other. T time units later, they are an additional Y units apart. What was the change in attractive force between these two planets during the time period T? G is gravitational force constant.

A. $GMN(1/x^2 - 1/y^2)$
B. $GMN(1/y^2)$
C. $GMN\{1/(x - y)^2\}$
D. $GMN\{1/x^2 - 1/(x + y)^2\}$

27. Torque is:

I. a scalar.
II. a vector.
III. force times lever arm.
IV. force times direction.
V. the perpendicular distance between the line along which the force acts and the axis of rotation of the rotating body.

A. I and III
B. II and III
C. I and IV
D. II and IV

SOLUTIONS

Force, Momentum, Circular Motion, and Gravitation

1. **C** To calculate the unknown force, use F = ma. Find the acceleration (a). Use the equation:

   $x = v_ot + \frac{1}{2} at^2$ to calculate a.

   $35 \text{ m} = (2\text{m/s})(5 \text{ sec}) + (\frac{1}{2} a)(5 \text{ sec})^2$

   $35 = 10 + (25/2)\, a$

   $50 = 25\, a \qquad a = 2 \text{ m/s}^2$

   Therefore, $F = (1 \text{ kg})(2 \text{ m/s}^2) = 2 \text{ N}$.

2. **A** Use the conservation of energy principle to solve this question. PE at maximum height = KE at contact with the ground.

   $mgh = \frac{1}{2} mv^2$

   Solving for v:

$v = (2\, gh)^{1/2}$

$v_{(10 \text{ g mass})} = [2\, g(2 \text{ m})]^{1/2} = (4\, g)^{1/2} = 2(g)^{1/2}$

$v_{(5 \text{ g mass})} = [2\, g(3 \text{ m})]^{1/2} = (6\, g)^{1/2}$

$v_{5 \text{ g}} - v_{10 \text{ g}} = (6\, g)^{1/2} - 2(g)^{1/2}$

3. **C** Because all the blocks are ultimately supported by the tension in T_1, the magnitude of T_1 is the sum of the weights of all three blocks, that is, 110 lb. The tension in T_2 is the sum of the weights of the lower two blocks, that is, 90 lb.

4. **D** During free fall, the man is weightless because the scale falls at the same rate as the man. Therefore, the man exerts no downward force on the scale.

5. **D** Solve this problem in two steps. The boy weighs 64 lb when the elevator is not moving; therefore, find his mass by dividing the acceleration of gravity (g) into the weight:

   $W = mg$

   $64 \text{ lb} = (m)(32 \text{ ft/s}^2)$

   $m = 2$

   When the elevator accelerates 4 ft/s^2 upward, it is accelerating "into" the boy (i.e., pushing up on the boy). The boy reciprocates by pushing back on the elevator with an equal and opposite force. The scale reads the net force with which the boy pushes down on the elevator.

   $F_{elevator} = F_{boy} = m(4 \text{ ft/s}^2)$

   $F_{gravity} = m(32 \text{ ft/s}^2)$

   Net force down $= m(4 \text{ ft/s}^2 + 32 \text{ ft/s}^2) = 2(36 \text{ ft/s}^2) = 72$ lb.

6. **B** When the elevator accelerates downward, it appears to lighten her weight because the acceleration of the elevator partly counteracts the acceleration due to the Earth's gravity. Look at the following diagram. Note these important terms in the diagram:

   $F_{elevator}$ = force on passenger owing to elevator accelerating downward

   $F_{passenger}$ = force of passenger on elevator, equal and opposite to $F_{elevator}$

   $F_{gravity}$ = force of gravity

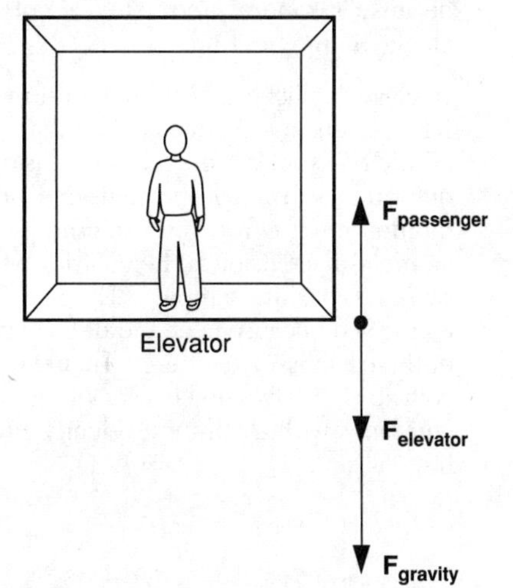

   A passenger presses downward on the floor of the elevator with the following net force:

   $F_{net} = F_{gravity} - F_{passenger}$

   Therefore, if the woman weighs 50 N when the elevator is moving downward, find her mass as follows:

$50\ N = F_{gravity} - F_{passenger}$

$50\ N = m(10\ m/s^2) - m(5\ m/s^2)$

$50\ N = m(5\ m/s^2)$

$m = 10\ kg$

Therefore, the woman's weight is $W = mg = (10\ kg)(10\ m/s^2) = 100\ N$.

7. **B** A steeper plane allows a mass to reach the bottom faster because the speed at which the blocks travel depends on the height of the plane. Recall that $v = (2\ gh)^{1/2}$. A 60-degree plane is taller than a 30-degree plane of equal length. Mass is not a factor.

8. **A** PE at the top of the hill = Work + KE at the bottom of the hill. The work term equals the heat dissipated by friction. Solving the preceding equation gives:

   $mgh = W - \frac{1}{2} mv^2$

   To solve this equation for W requires the mass of the sled, the height of the hill, and the final velocity.

9. **D** Impulse is defined as the product of the average force acting on an object and the time interval (Δt) over which the force acts. Impulse is a vector quantity. The term mv, which is the product of mass and velocity, is the definition of momentum. Choice I can be excluded because it is merely momentum, not change in momentum.

10. **C** In elastic collisions, both momentum and KE are conserved. An example of an elastic collision is the head-on collision of billiard balls. In inelastic collisions, only momentum is conserved. An example of an inelastic collision is the collision of two cars. In this example, KE is not conserved because heat and deformation of the cars dissipate energy. In inelastic collisions, the two objects stick together and move with a common velocity after the collision.

11. **B** To solve this problem, use conservation of momentum.

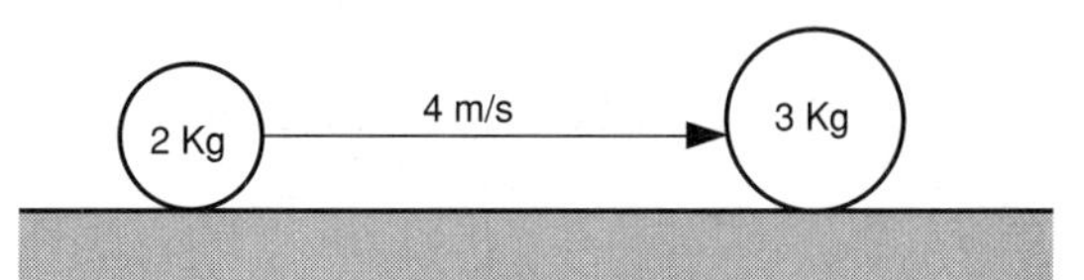

Before collision

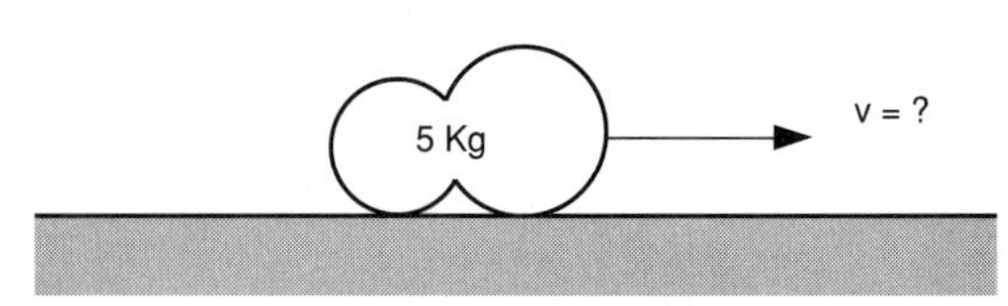

After collision

Remember that in inelastic collisions, the two objects mesh together after the collision.

$mv_i + mv_i = m_f v_f$

$(2\ kg)(4\ m/s) + (3\ kg)(0\ m/s) = (5\ kg)(v_f)$

$v_f = \frac{8}{5}\ m/s$

12. **D** The KE of the system $= \frac{1}{2}mv^2$. The mass of the system is equal to the sum of the masses of the blocks (they stick together in inelastic collisions).

    Therefore, $KE = \frac{1}{2}(14\ kg)(10\ m/s^2) = 700\ J$.

13. **A** The best way to solve the problem is to draw a momentum diagram.

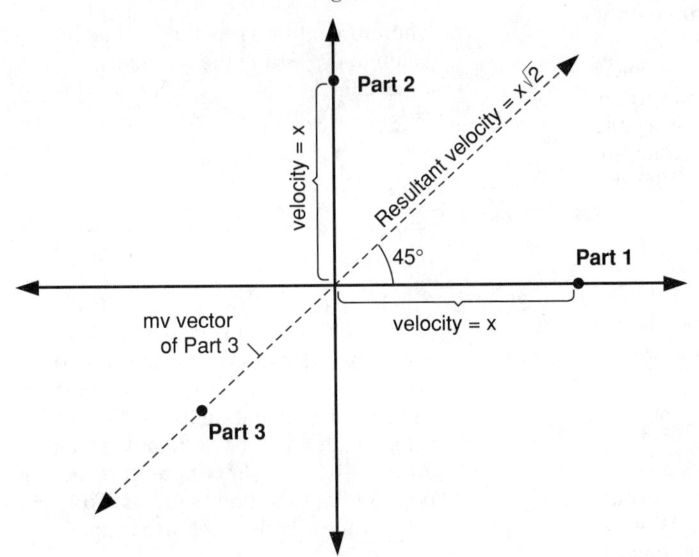

Note that two of the parts are perpendicular to each other. In addition, note that the resultant vector, determined by adding the momentum vectors of parts 1 and 2, is equidistant and midway between them. Because the initial momentum of the block is zero ($v_i = 0$), the sum of the momentum of parts 1, 2, and 3 must also equal 0 (conservation of momentum). To get this 0, part 3 must have a momentum vector directed 180° opposite the result of parts 1 and 2. This momentum vector has the same velocity ($x\sqrt{2}$), but a different direction from the resultant vector of parts 1 and 2.

14. **D** Choices A–C are all true. Elastic collisions conserve momentum, KE, and do not mesh after the collision. They obey Newton's third law; that is, objects impart an equal and opposite reaction force to one another during the collision.

15. **B** Use conservation of energy to solve this problem. PE gained after bullet impact equals the KE of the bullet–block system at impact. The PE gained (vertical swing) = mgh, where m = total system mass and h = height gained by the system. Thus, PE = (x + y)gh = KE of the bullet–block system after impact.

16. **B** In this inelastic collision problem, use the conservation of momentum to solve:

$$(mv_i)_{arrow} + (mv_i)_{block} = (mv)_{arrow\text{–}block}$$

$$(0.1\ kg)(100\ m/s) + (0.9\ kg)(0) = (1\ kg)(x)$$

$$x = 10\ m/s$$

17. **A** KE and momentum are the quantities that are conserved. Choice C is incorrect because momentum is shared proportionally between the blocks, depending on the relative mass and velocity. Choice D is not correct because PE is not conserved.

18. **C** Study the following diagram.

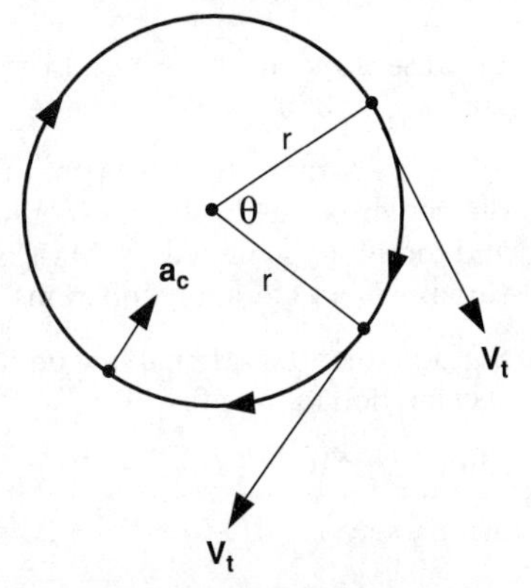

The diagram shows that a tangential velocity exists, because velocity is defined as a vector quantity having both a magnitude and a direction. Even though the object rotates at constant speed, it constantly changes direction, and therefore has a tangential velocity (v_t). Angular velocity, or ω, is defined as ($\Delta\theta/\Delta t$) or v_t/r.

Since the object has a v_t and a radius r, it has a ω value. If an object is traveling in a circular path, a centripetal force must be acting on the object. Objects in circular motion always have an a_c, and always have an ω. An applied inward, or centripetal, acceleration (a_c), associated with a centripetal force, is necessary to prevent the rotating object from leaving the circular path.

$a_c = (v_t)^2/r = r\omega^2$

Centripetal force is a product of the centripetal acceleration and the mass of the body. No tangential acceleration exists at constant speed.

$F_c = ma_c = mv^2/r = mr\omega^2$

19. **A** Remember that ideally, at constant speed, constant velocity is not necessary (direction changes). However, if constant speed is assumed, there would be no a_t, because a_t depends on changes in v_t. Because $\Delta v_t = 0$ in constant velocity situations, $a_t = 0$. Because a_c still exists, the net acceleration on the object is only centripetal.

20. **B** $\omega = \Delta\theta/\Delta t$ or v_t/r. $\Delta\theta/\Delta t = (10\pi$ radians)/5 sec $= 2\pi$ rad/s.

21. **C** $s = r\theta$, where s = arc length (distance), r = radius, and θ = angle.

$10\pi = \frac{1}{2}\theta \qquad 10\pi/(\frac{1}{2}) = \theta$
$\theta = 20\pi$ radians

Because π radians $= 180°$, $20\pi = 20(180°) = 3600°$.

22. **C** $a_t = r\alpha$. As the point gets farther from the center of the flywheel, r increases, and therefore, a_t increases. Mass is not a variable in acceleration equations.

23. **A** Suppose that the original tangential acceleration (a_t) = 10.

After 1 sec, $10 - 1/2(10) = 10 - 5 = 5$

After 2 sec, $5 - 1/2(5) = 5 - 2.5 = 2.5$

After 3 sec, $2.5 - 1/2(2.5) = 2.5 - 1.25 = 1.25$

Therefore, after 3 sec, the a_t value has fallen to (1.25 to 1) the original a_t. Rearranging, this gives: $1.25/10 = (\frac{5}{4})(\frac{1}{10}) = \frac{1}{8}$ or $0.125\ a_t$.

24. **D** The number of rotations $= \omega t = (5$ rad/sec)(5 sec) = 25 rad.

25 rad(1 rotation/2π rad) = $25/2\pi$ rotations.

25. **C** The force of attraction between two bodies is $F = Gm_1m_2/r^2$, where G is the universal gravitation constant, m_1 and m_2 are the masses of the two bodies, and r is the distance separating them. The g value of a body (acceleration due to gravity) $= Gm_{body}/r^2$, where m = mass of body and r = the body's radius. To compare two bodies, set up a ratio:

$$\frac{\textbf{Planet X} = g_x =}{\textbf{Earth} = g_e =} \frac{\frac{Gm}{r^2}}{\frac{Gm}{r^2}} = \frac{\frac{1/2\ m_e}{(1/4\ r_e)^2}}{\frac{m_e}{(r_e)^2}} = \frac{\frac{1/2\ m_e}{1/16\ r_e^2}}{\frac{m_e}{r_e^2}} = 8$$

26. **D** To solve the problem, find the force between the planets at both times.

$F = Gm_1m_2/r^2$

Originally, $F_i = GMN/X^2$ (stronger attractive force).

T units later, $F_f = GMN(X + Y)^2$ (the planets are farther apart and therefore have a weaker attractive force). The difference between the attractive force of F_i and F_f is:

$F_i - F_f = GMN/X^2 - GMN/(X + Y)^2 = GMN[1/X^2 - 1/(X + Y)^2]$

27. **B** Statements II and III accurately describe torque. Because torque is a vector quantity, statement I is incorrect. Statements IV and V are also incorrect and should be eliminated.

HIGH-YIELD REVIEW QUESTIONS

Section I: Physics

Rotational Dynamics, Mechanical Properties, and Fluids

1. The moment of inertia (I) is:

I. dependent on the mass distribution and position of the axis of rotation.
II. $I = mr^2$ for a point mass rotating about a fixed axis.
III. needed to calculate the angular momentum of shapes other than a point mass.
IV. greater for hollow objects than solid objects of the same shape, mass, and size.

A. I, II, and IV
B. II, III, and IV
C. I, II, and III
D. I, II, III, and IV

2. When an ice-skater spins and draws in her arms, her angular velocity increases. This increase occurs because:

A. her I value increases and angular momentum is conserved.
B. her I value decreases and angular momentum is conserved.
C. her I value increases and linear momentum is conserved.
D. her I value decreases and linear momentum and centripetal force are conserved.

3. A block is swung in a horizontal circle of radius r, at constant angular velocity. How do linear and angular momentum compare?

A. Linear and angular momentum are constant.
B. Linear and angular momentum are changing.
C. Linear momentum changes; angular momentum is constant.
D. Angular momentum changes; linear momentum is constant.

4. The unit of the moment of inertia can be:

A. kgm^2.
B. kg/s^2.
C. kgm/s^2.
D. kgms.

5. The terms stress and strain for the length of an object define two measurements of elasticity. The definitions of stress and strain are:

I. tensile force/area.
II. tensile force/unstressed length.
III. elongation/area.
IV. elongation/unstressed length.

A. I and II
B. II and III
C. III and IV
D. I and IV

6. The unit of an elastic modulus can be:

A. N/m^2.
B. kgm/s^2.
C. $(N)(s)/m^2$.
D. $(N)(kg)/ms^2$.

7. Consider a cylinder of height H, with an open top, and filled with a liquid of density D. Atmospheric pressure is P, and the surface area of the top of the cylinder is S. What is the force acting over the top surface of the liquid?

A. PH
B. PS
C. PDH
D. PDSH

8. What is the pressure at point A? Assume the density of the liquid is D and $P_{atm} = P$. Ignore units.

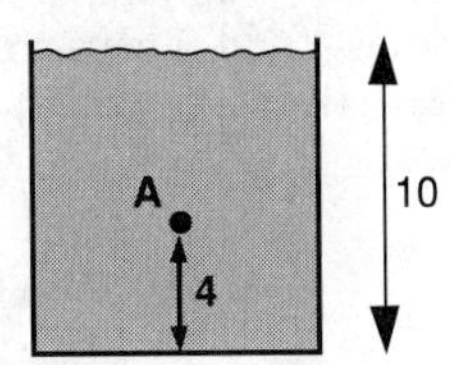

A. P + 4Dg
B. 6Dg
C. 4Dg + 6PDg
D. P + 6Dg

9. Two objects, A and B, are submerged in water. What is the pressure difference between A and B?
water = 1000 kg/m^3 g = 10 m/s^2

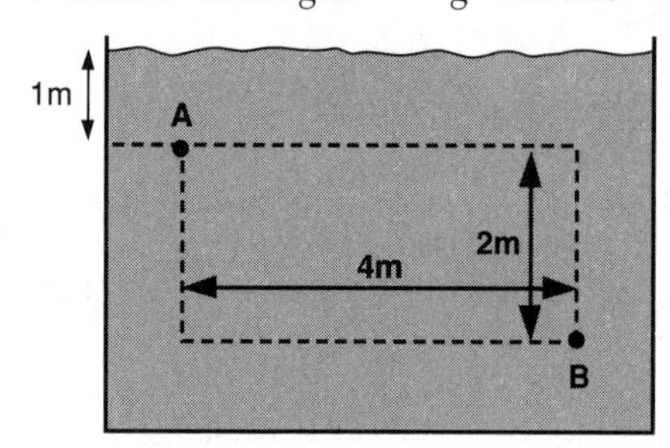

A. 20,000 kgm/s^2
B. 40,000 kgm/s^2
C. 30,000 kgm/s^2
D. 60,000 kgm/s^2

10. Consider the following diagram. Points A, B, and C are collinear and parallel to the base of the tank. They represent points in a fluid filled tank. How do the pressures at A, B, and C compare?

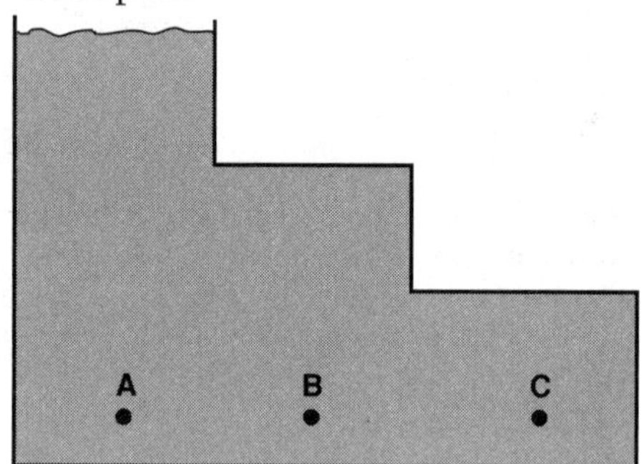

A. A > B > C.
B. C > B > A.
C. All have the same pressure.
D. There is not enough information for comparison.

11. Consider the hydraulic lift depicted. Suppose that a force of 10 N is applied downward on piston B. Pistons A and B have areas of 2 m^2 and 6 m^2, respectively. What upward force is exerted at A?

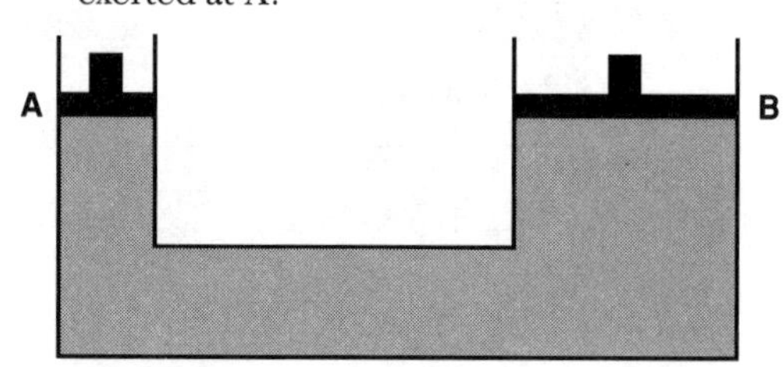

A. $10/3$ N
B. 15 N
C. 30 N
D. 60 N

12. A metal sphere with a volume of 1 m^3 is submerged in a tank of water. What buoyant force acts on the sphere?
water = 1000 kg/m^3 g = 10 m/s^2

A. 1000 N
B. 100 N
C. 10,000 N
D. Not enough information to determine

13. A metal cube is suspended in a pail of water by a string. The pressue is greatest against the:

A. side.
B. top.
C. bottom.
D. all sides equally.

14. An object of uniform density floats on water with 60% of its volume submerged. What is its specific gravity?

A. 0.6
B. 0.4
C. 1.0
D. 0.25

15. An object floats in a liquid of specific gravity = X, with 25% of the object floating above the level of the liquid. What is the specific gravity of the object compared to that of the liquid?

A. 1.0 X
B. 0.375 X
C. 0.25 X
D. 0.75 X

16. A block weighs 5 N; but when submerged in water, it weighs 2 N. What is the specific gravity of the block?

A. 3
B. $5/3$
C. 7
D. $3/2$

17. An object of unknown volume is submerged in a liquid of density 100 kg/m^3. The object weighs 10 N in air, but ony 4 N when submerged in the liquid. What is the volume of the object?   g = 10 m/s^2

A. 0.02 m^3
B. 0.006 m^3
C. 0.40 m^3
D. 0.007 m^3

18. Fluid flows from left to right in this pipe. Where is the greatest pressure?

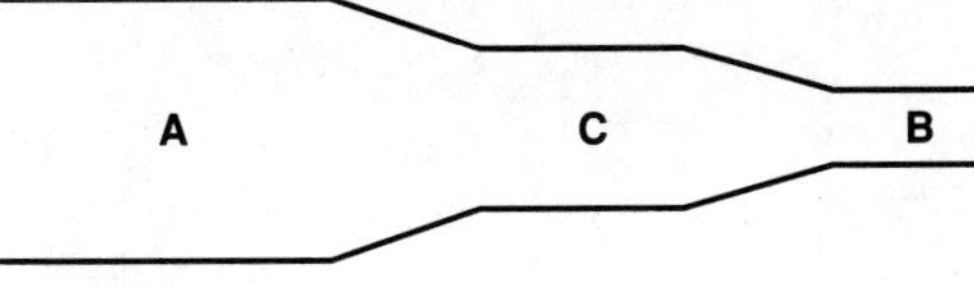

A. A
B. B
C. C
D. Same in all sections

19. A pipe of constant diameter d has three regions, A, B, and C, through which fluid can freely flow. Fluid flows from left to right. Which region has the greatest pressure?

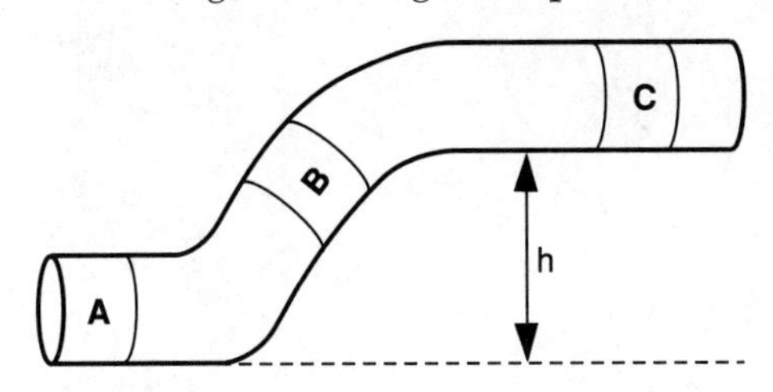

A. A
B. B
C. C
D. Same pressure for all

20. Fluid flows from left to right in a pipe system such that the velocity at A is four times the velocity at B. What is the ratio of the diameter at A to the diameter at B?

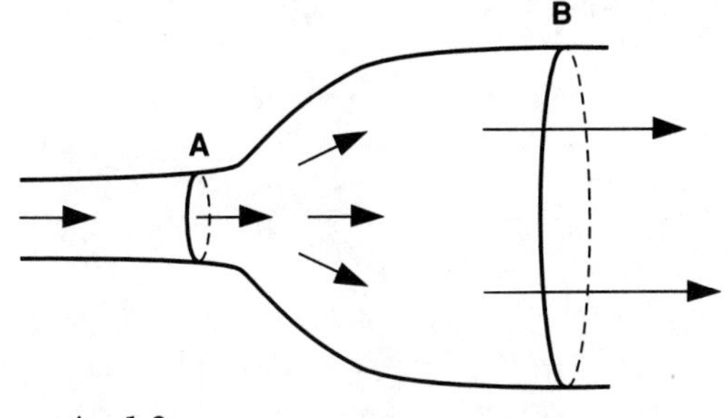

A. 1:2
B. 1:4
C. 1:3
D. 1:1.5

21. A large water tower develops a leak 20 meters from the surface of the water. What is the velocity of the water exiting the leak? Assume g = 10 m/s^2.

A. 40 m/s
B. 30 m/s
C. 20 m/s
D. 10 m/s

22. Fluid flows from left to right in the following pipe system. Three vertical spouts allow fluid under enough pressure to spray out of the spouts. Assuming that all the spouts spray out liquid, which spout sprays liquid highest into the air?

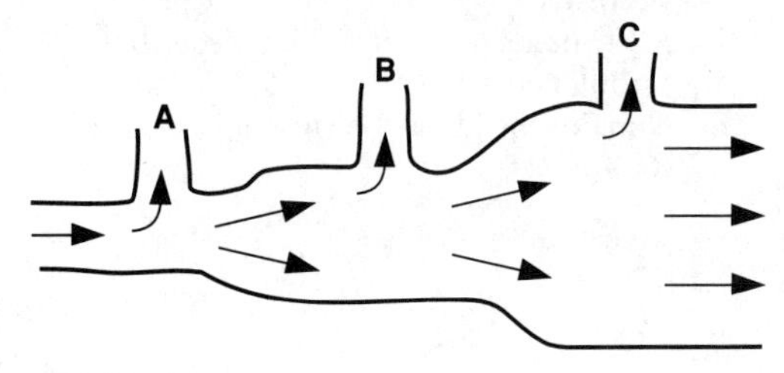

A. A
B. B
C. C
D. All spray heights equal

23. Surface tension:

I. results from adhesion forces.
II. results from cohesion forces.
III. is involved in capillary action.
IV. varies indirectly with surface area.

A. I, II, and IV
B. I and III
C. II and III
D. II, III, and IV

24. A drop of water sits on a piece of wax paper. Which force keeps the droplet spherical and in minimal contact with the wax?

I. Adhesion
II. Cohesion
III. Hydrophobic interactions
IV. Capillary action
V. Surface tension

A. I, II, IV, and V
B. I, III, and V
C. II, III, and V
D. I, II, III, IV, and V

25. The viscosity of a fluid:

I. opposes fluid flow.
II. depends on the distance between fluid layers.
III. depends on shear stress.
IV. depends on turbulent flow of fluids.
V. is given in units of pascal second.

A. I, II, IV, and V
B. I, III, IV, and V
C. I, II, III, IV, and V
D. I, II, III, and V

26. Which statement does NOT describe turbulent fluid flow?

A. When flow rate exceeds a critical velocity, laminar flow can become turbulent.
B. Turbulence varies indirectly with viscosity.
C. A Reynolds number > 2000 describes turbulence.
D. None of the above because all are true statements.

SOLUTIONS

Rotational Dynamics, Mechanical Properties, and Fluids

1. **D** Statements I–IV are true. Statements I and II describe the moment of inertia. It can be a distribution of mass over space. The moment of inertia (I) is substituted for mass in the equation for torque: $\tau = I\alpha$; angular momentum: $L = I\omega$; KE of rotation: $\frac{1}{2}I\omega^2$, and other formulas of rotational dynamics. Generally, the I value for hollow objects is greater than the I value for solid objects of the same size and shape because more torque is needed to give a hollow object an angular acceleration in that more mass is located farther from the axis of rotation, and $I = mr^2$. Therefore, a cylindrical shell has a greater I value than an identical solid cylinder if the masses of the two objects are equal.

2. **B** Angular momentum ($L = I\omega$), a conserved quantity. When the woman spins with her arms outstretched, L is equivalent to the momentum when she draws in her arms. $I\omega$ (arms outstretched) = $I\omega$ (arms drawn in). However, drawing in the arms decreases I and therefore increases ω. The product of $I\omega$, nevertheless, stays the same.

3. **C** Linear momentum = mv. Because the tangential velocity of the block changes at any new instant of time (due to vector direction changing), the product mv changes. Because angular momentum = $I\omega$ and ω is constant, one can conclude that I is constant (the orientation and position of the block to the axis or rotation is not changing), and the product $I\omega$ is constant.

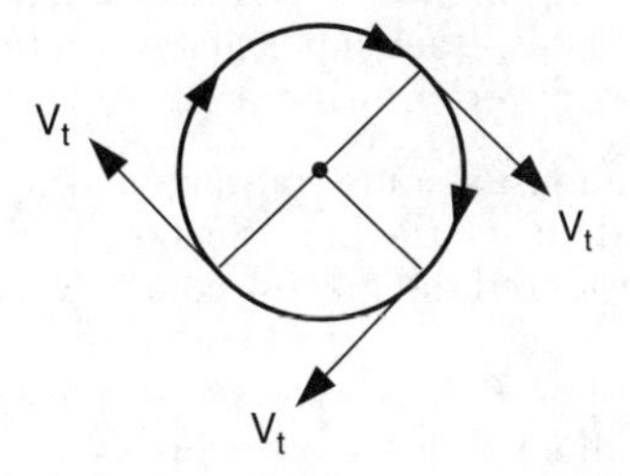

4. **A** Because $I = mr^2$, the units can be $(kg)(m^2)$.

5. **D** Stress = F/A. Strain = $\Delta L/L_o$. Therefore statements I and IV are correct and the best answer is choice D. Note that stress has the same units as pressure.

6. **A** The elastic modulus is defined as:

   modulus = stress/strain = Y(modulus) = $(F/A)/(\Delta L/L_o)$

   Force is in newtons, area is in m^2, and both ΔL and L_o are in meters.

   Therefore, $Y = (N/m^2)/(m/m) = N/m^2$.

7. **B** Force acting over the top surface of a liquid = (atmospheric pressure)(surface area of cylinder top) = PS.

8. **D** The equation needed to solve this problem is: $P = P_{atm} + \rho gh$, where h = depth below the surface of the liquid. The equation makes conceptual sense. The pressure of an object below the surface of a liquid is the sum of the atmospheric pressure acting on the surface of the fluid and the additional pressure (weight) of the fluid above the submerged object.

   Therefore, in this problem: $P = P + Dg(6) = P + 6Dg$.

9. **A** The difference in pressure between two points $\Delta P = \rho g \Delta h$, where ρ is the density of the fluid, and Δh is the vertical difference between the two points.

   Therefore, in this problem: $\Delta P = (1000\ kg/m^3)(10\ m/s^2)(2\ m) = 20{,}000\ kgm/s^2$

10. **C** Pascal's law says that in equilibrium conditions (i.e., still fluid, v = 0), the pressure at one point in a liquid is transmitted uniformly to all parts of the liquid at the same depth. Even though more water lies over point A than over point B or C, they are all at the same pressure. This statement must be true, or fluid would flow from one of these points to the others at lower pressure.

11. **A** Remember that an incompressible fluid is found in the hydraulic lift that transmits pressure from one side of the lift to the other. Pascal's law says that the same pressure occurs throughout the fluid. Solve this problem by remembering that force per unit area (pressure) on both sides of the hydraulic system is equal:

$F_B/A_B = F_A/A_A$

Substituting the numbers given:

$10\ N/6m^2 = F_A/2m^2$

$F_A = 10/3\ N$

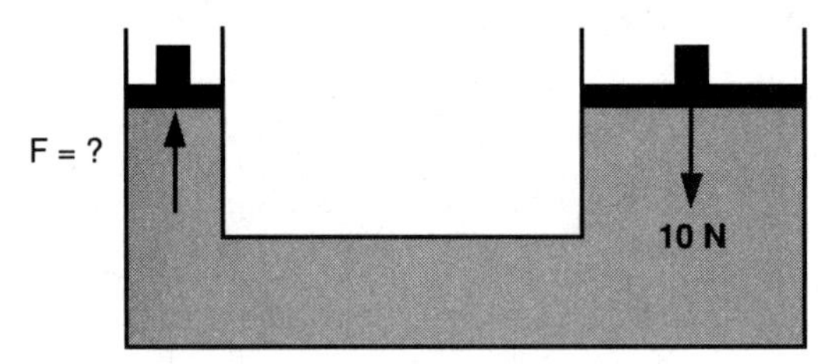

Notice that a small force (and surface area) at A is associated with a large force and surface area at B.

12. **C** Archimedes' principle says that a body submerged or incompletely submerged in a fluid is buoyed up by a force equal to the weight of the fluid that it displaces. Thus, B = (V)(ρ)(g), where B = buoyant force, V = volume of fluid displaced, and ρ = density of the liquid. In this problem, B = (1 m^3)(1000 kg/m^3)(10 m/s^2) = 10,000 N.

13. **C** The pressure is greatest on the side at the greatest depth, that is, the bottom. Remember that $P = P_{atm} + \rho gh$, where h = depth.

14. **A** Following are the two basic methods to solve this problem.

Method 1: A manipulation of the definition of specific gravity says that the *percentage of the object that is submerged in water equals the specific gravity of the object.* Therefore, the specific gravity must be 0.6 because 60% of the object is submerged.

Method 2: This method requires calculations. Suppose an object has a volume of 1 m^3. If it floats with 0.6 under water, this amount corresponds to a displaced weight of water: B = Vρg = (0.6 m^3)(1000 kg/m^3)(10 m/s^2) = 6000 N. By comparison, the whole object would displace: (1 m^3)(1000 kg/m^3)(10 m/s^2) = 10,000 N or water. Therefore, the specific gravity = (6000 N)/(10,000 N) = 0.6.

15. **D** Question 14 outlined a way to predict the specific gravity based on the percentage of the object that was submerged. The percentage submerged = 75% of the volume of the object. This percentage corresponds to the relationship between the specific gravity of liquid X and the object.

16. **B** The specific gravity equation is as follows:

Specific gravity (SG) = $W_o/(W_o - W_a)$, where W_o is the actual object weight and W_a is the apparent weight of the object in liquid.

Solve by plugging in the numbers given in the question: SG = 5 N/(5 N − 2 N) = 5/3.

17. **B** A good way to find the volume of an object submerged in a liquid is to use the following formula:

Volume of object: $(W_o - W_a)/\rho g$

Using the numbers given: Volume of object = (10 N − 4 N)/(100 kg/m^3)(10 m/s^2).

Because 1 N = 1 kgm/s^2, (6 kgm/s^2)/(1000 kg/m^2s^2) = 0.006 m^3.

18. **A** Bernoulli's equation says that in a system, $P + \rho gh + \frac{1}{2}\rho v^2$ = a constant; that is, the sum of the terms in sections A, B, and C is constant. In pipes of small radii, fluid flows faster than pipes of greater radii; that is, the $\frac{1}{2}\rho v^2$ term is greatest for pipe B, with C next and A the least. The ρgh term is identical for all three sections because h (the relative height) is the same for all three sections. Because all three sections must equal the same constant in Bernoulli's equation, the P term in A must be greater than C, and even greater than B. The pressure term (P) increases as cross-sectional area increases. Remember that when fluid flows faster, pressure drops.

19. **A** One must assume constant velocity of fluid flow in the pipe. Section A can be considered the "lowest point." Using Bernoulli's equation for all three sections, $P + \rho gh + \frac{1}{2}\rho v^2$ = constant, and the ρgh term for section A is lower than B or C, (smaller h value for section A). Therefore, the pressure of A is the greatest.

20. **B** The continuity equation states: $A_aV_a = A_bV_b$ in a freely flowing system. If $V_a = 4V_b$, $V_a = x$ and $V_b = 4x$.

$A_ax = A_b(4x)$ $\quad A_a/A_b = x/4x = ¼$

21. **C** The speed with which the water exits the leak is equivalent to the speed that an object attains falling 20 meters. Using conservation of energy: $mgh = ½ mv^2$, derive Toricelli's theorem: $v = (2gh)^{1/2}$ $v = [(2)(10)(20)]^{1/2} = 20$ m/s.

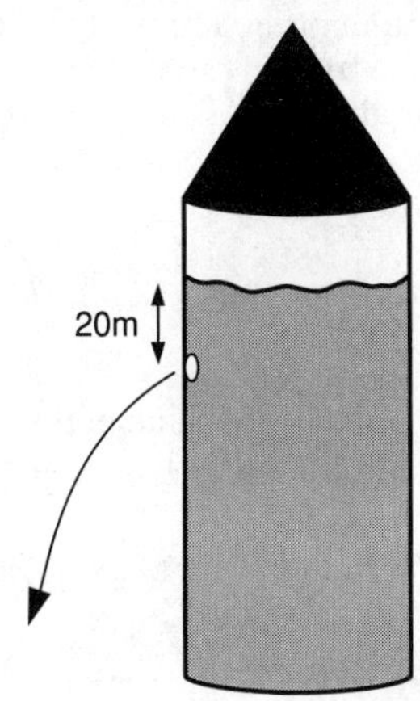

22. **C** In this problem, use Bernoulli's equation: $P + \rho gh + ½ \rho v^2 =$ a constant. A vertical pressure gradient occurs at any point in the pipe. The height difference between A and C is overcome only for sufficiently large flow rates. Because narrow pipes have higher rates of flow than wide pipes, Bernoulli's equation shows that pressure is lower in pipes with narrow diameters than in pipes with larger diameters. The highest pressure is in pipe C and allows pipe C to spray water the highest through the release spout.

23. **D** Surface tension of a liquid is defined as the increase in energy of the liquid per unit increase in surface area. $S = \Delta E/\Delta A$. Surface tension is caused by cohesive attraction of molecules (individual molecules of the same type attract each other). Adhesion, on the other hand, occurs when liquid molecules adhere to molecules of a solid or molecules of a different liquid. If adhesion is greater than cohesion, the liquid wets the surface of the solid. Both adhesion and surface tension are involved in capillary action, the rise of a liquid column in a narrow tube.

24. **C** Cohesive forces act to pull the water molecules together. Hydrophobic interactions occur between polar water molecules and nonpolar wax molecules. Surface tension also pulls the droplet into a spherical shape. Adhesion does not play a major role here because little attraction occurs between water molecules and wax molecules. Capillary action refers to a narrow column of water moving up a tube.

25. **D** Viscosity is defined as the shear stress needed to maintain the laminar flow of two planes of liquid separated by a distance (l) and moving with velocity (v). The viscosity coefficient $= \eta = (F/A)/(v/l)$, where F = force, A = area, l = distance between layers of fluid moving with velocity v in relation to one another, with area A, and with an applied force of F.

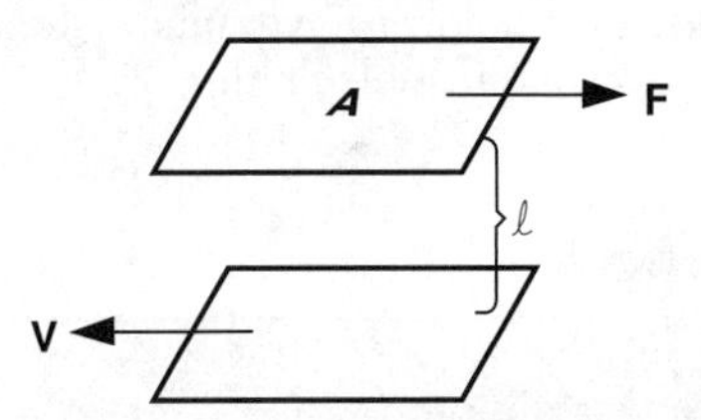

26. **D** Choices A–C describe turbulent fluid flow. Turbulence can be measured using the Reynolds number defined by: $R = dvr/\eta$, where d = density of fluid, v = velocity of fluid, r = radius of pipe, and η = fluid viscosity. An R greater than 2000 denotes turbulent flow.

HIGH-YIELD REVIEW QUESTIONS

Section I: Physics

Comprehensive Quiz

1. What is the ratio of "g" on planet X compared to planet Y?

Data Available
Mass of X = a
Mass of Y = 2a
Radius of X = b
Radius of Y = 2b

A. 1:2
B. 2:1
C. 1:4
D. 4:1

2. A 2-kg metal sphere rotates with an angular velocity of 5 radians/sec in a circular path with a 2-meter radius. What is its acceleration toward the center of the circular path?

A. 100-kg rad^2/s^2
B. 100 N rad^2
C. 50 $rad^2 m/s^2$
D. 20-kg rad^2/s^2

3. Consider the following hydraulic system filled with an incompressible fluid:

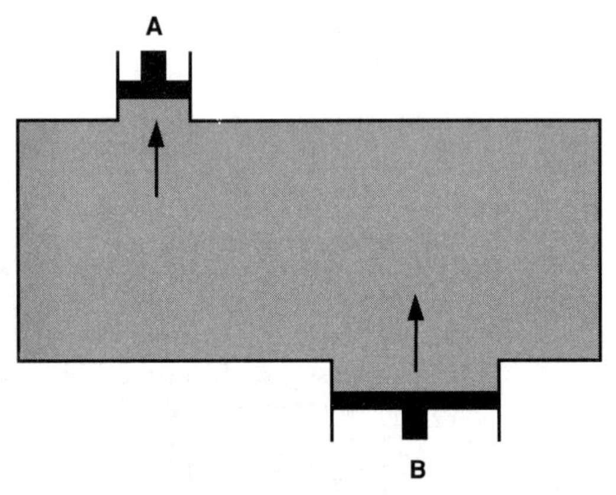

Suppose a force X is applied to piston B of surface area Y m^2. What can maximize the force exerted on an object resting on top of piston A?

A. A large surface area of piston A
B. A small surface area of piston A
C. A surface area of piston B larger than Y
D. Two of the above are correct.

4. A 1-kg block slides down a ramp that is 2 meters high. Its velocity at the bottom of the ramp is 4 m/s, and it takes 10 seconds to reach the bottom of the ramp. How much power does the friction associated with this movement produce? Assume g = 10 m/s^2.

A. 2.0 W
B. 1.8 W
C. 1.5 W
D. 1.2 W

5. What is the approximate value of the unknown weight W, using the accompanying diagram?

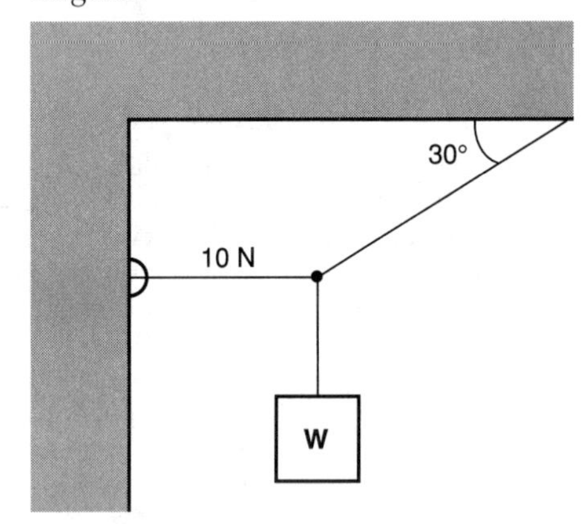

A. 5.75 N
B. 11.5 N
C. 7.5 N
D. Not enough information provided

6. A 30-kg block is pulled along the floor by a 200-N tension force directed at 37 degrees above the horizontal. If the coefficient of kinetic friction is 0.3, what is the acceleration of the block? Assume $g = 10\ m/s^2$.

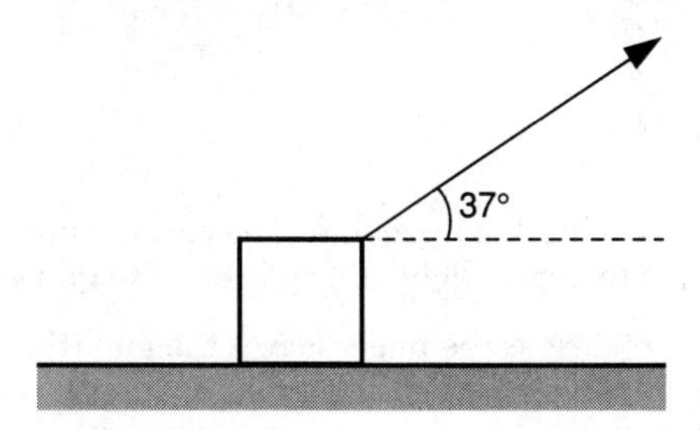

A. $3.23\ m/s^2$
B. $2.33\ m/s^2$
C. $2.57\ m/s^2$
D. $3.53\ m/s^2$

7. Three blocks of mass 6 kg, 5 kg, and 4 kg, respectively, are shown next, resting on one another and a tabletop.

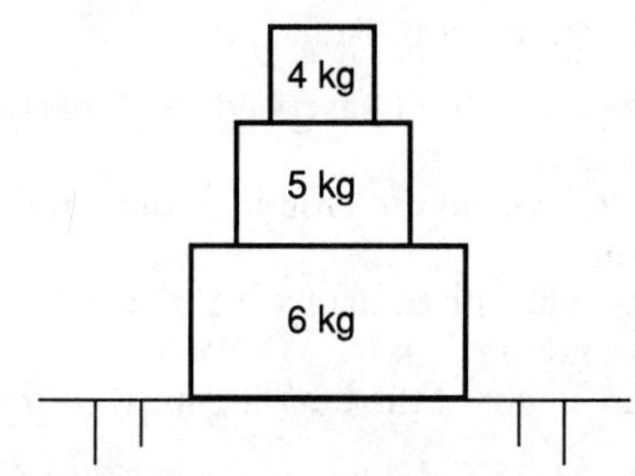

What force, if any, does the table exert on the 6-kg block? Assume $g = 10\ m/s^2$.

A. 150 N
B. 60 N
C. 15 N
D. 0 N

8. A man lifts and holds a 10-kg box ½ meter over his head. He walks with the block a distance of 5 meters with the block over his head. How much work does he do in this process? Assume that the man is 2 meters tall and he picks up the box from the floor. Ignore any work done in accelerating the box from rest to walking speed. Assume $g = 10\ m/s^2$.

A. 250 J
B. 750 J
C. 300 J
D. 800 J

9. A block of mass M rests on a surface. A string is attached to the block and exerts a tension T on the block. A friction force, F, acts to keep the block at rest. The coefficient of this static friction is 10.0. What is the tension T given by?

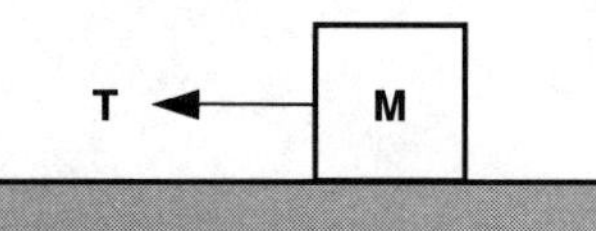

A. T = 10 Mg
B. T = Mg/10
C. T = 10 g
D. None of the above

10. The following object has a coordinate axis system as shown, with the origin at point X. The portion of the object to the right of point X has a center of gravity at R, and a mass of 100 grams. The portion to the left of X has a center of gravity at T and a mass of 70 grams. What is the distance between X and the center of gravity for the object?

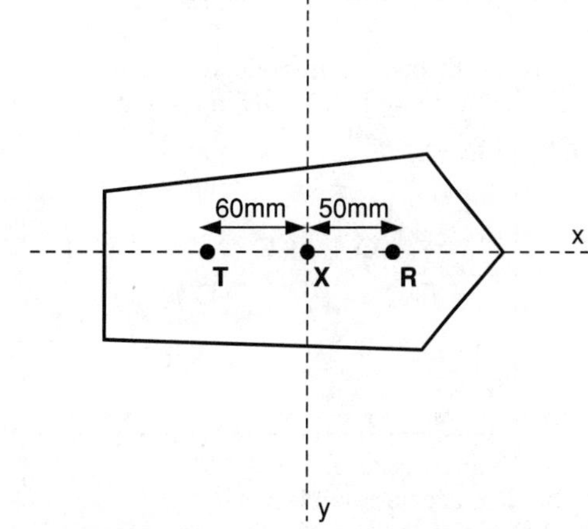

A. 3.2 mm
B. 7.6 mm
C. 4.7 mm
D. 5.9 mm

11. A projectile is shot at a 45-degree angle to the horizontal, from ground level. What data are needed to calculate its time of travel to hit the ground?

A. v_o and g
B. v_o, v_f, and g
C. v_o, weight, and g
D. No additional information

12. A tin rod is subjected to strain. Its original length is 5 meters; but after force is applied, it measures 5.5 meters. What tensile strain acts on the rod?

A. 0.1
B. 0.5
C. 10.5
D. 15.5

13. A car with initial velocity of 6 m/s accelerates at 2 m/s^2 for 3 seconds. How far does the car travel in 3 seconds?

A. 27 meters
B. 30 meters
C. 21 meters
D. 40 meters

14. Fluid flows horizontally from a pipe of narrow to wide diameter. In this process:

A. water pressure drops.
B. water pressure increases.
C. there is no difference in water pressure.
D. there is not enough information.

15. A ball is thrown in the trajectory shown. At the highest point vertically in the path, which statement is true?

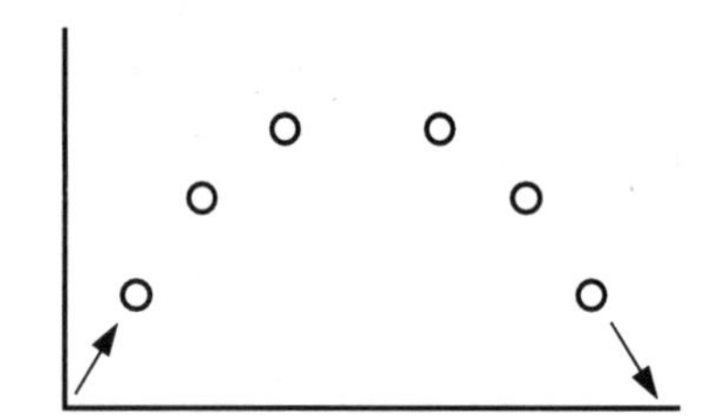

A. Velocity is zero.
B. The acceleration is zero.
C. Velocity and acceleration are perpendicular.
D. Velocity and acceleration are parallel and both equal.

16. An object is dropped from rest. It falls a distance of y_1 the 1st second and y_2 the 2nd second. What is the ratio of y_2/y_1? Assume $g = 10$ m/s^2.

A. 1:1
B. 2:1
C. 3:1
D. 4:1

17. An unknown substance is found in cubical form with an edge length of 2 cm. It has a mass of 16 grams. What is its specific gravity? Assume water = 1 g/cm^3.

A. 0.5
B. 1.0
C. 1.5
D. 2.0

18. A gun is fired in a vacuum. The acceleration of the bullet in flight after it leaves the gun is:

A. greatest as the bullet leaves the gun (the gun itself has finished accelerating the bullet).
B. decreased as the bullet approaches the target.
C. the same during the entire flight.
D. two of the above.

19. Two balls are projected horizontally from a tall building at the same instant, one with speed "x" and the other with speed "½ x". Which statement is true?

A. The ball with initial speed "½ x" reaches the ground first.
B. Both balls hit the ground at the same time.
C. The ball with initial speed "x" reaches the ground first.
D. The height of the building must be given.

20. A man stands on a frictionless surface and throws a ball of mass ¼ M to the right with velocity V. The man has mass 12 M. What is the man's velocity?

A. 48 V to the right
B. 3 V to the right
C. $\frac{1}{24}$ V to the left
D. $\frac{1}{48}$ V to the left

21. If acceleration is zero, which statement is true?

A. The velocity can be constant.
B. The velocity can be zero.
C. Both of the above
D. Neither of the above

22. The following curve shows the displacement of an object moving along a straight line as a function of time. Which curve shows the velocity of the object with respect to time?

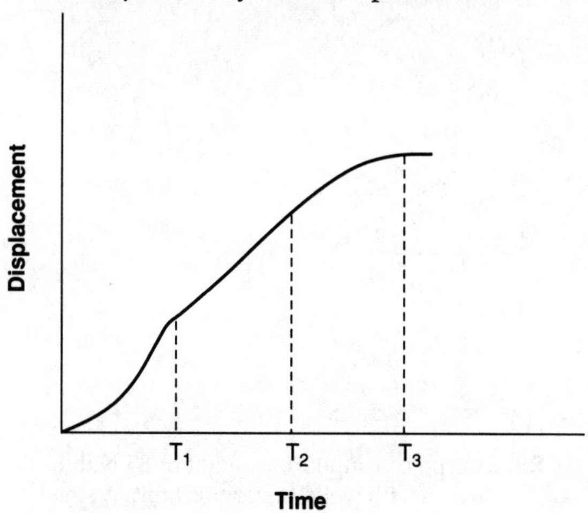

A.
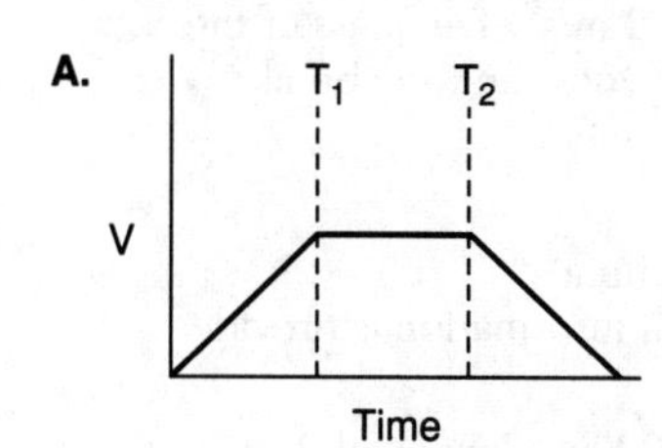

B.
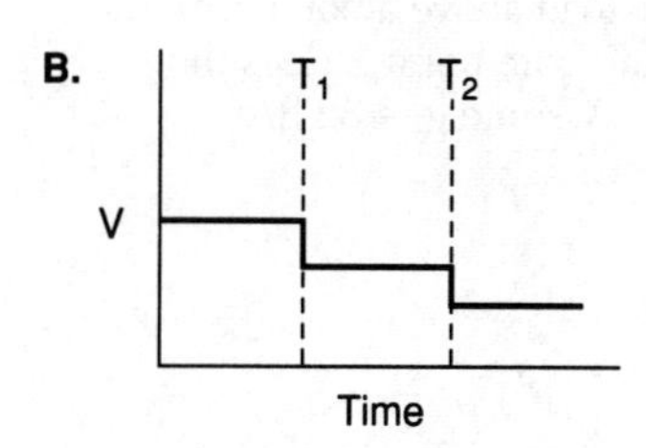

C.
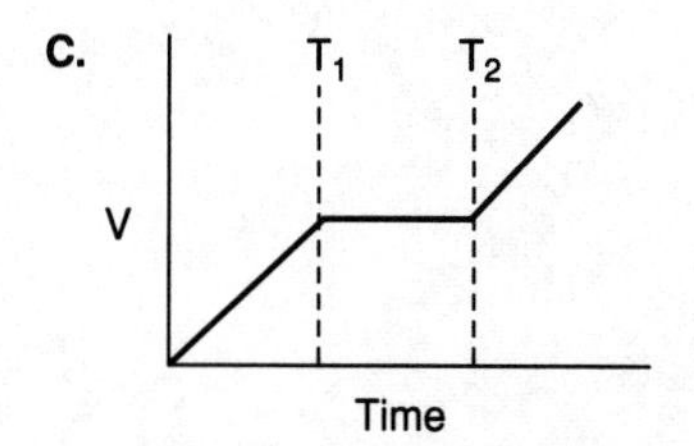

23. Speed describes:

A. the velocity of an accelerated body.
B. the magnitude of acceleration.
C. the velocity of a body with constant, but unknown direction.
D. the magnitude of velocity.

24. The resultant addition product of vectors A and B is closest to:

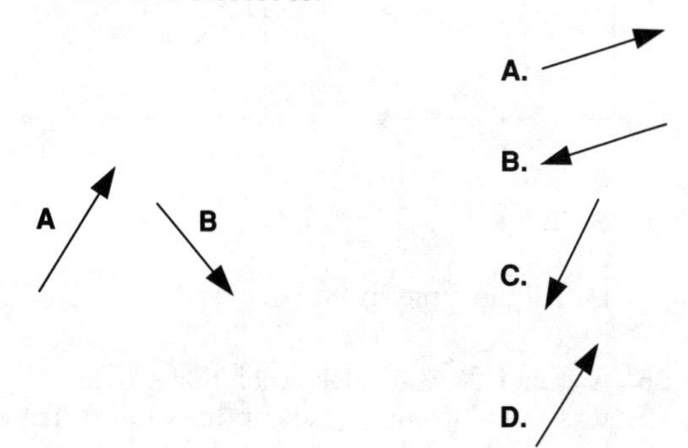

25. Which block has the greatest speed at point X? Assume that the tracks are frictionless and the blocks are of identical mass. Begin from rest at the same vertical height of 5 meters.

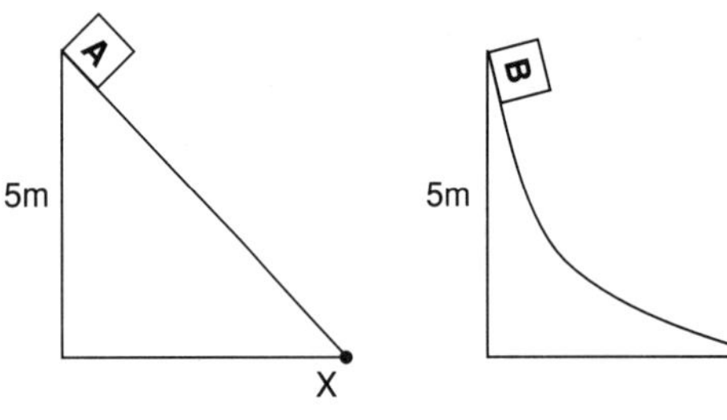

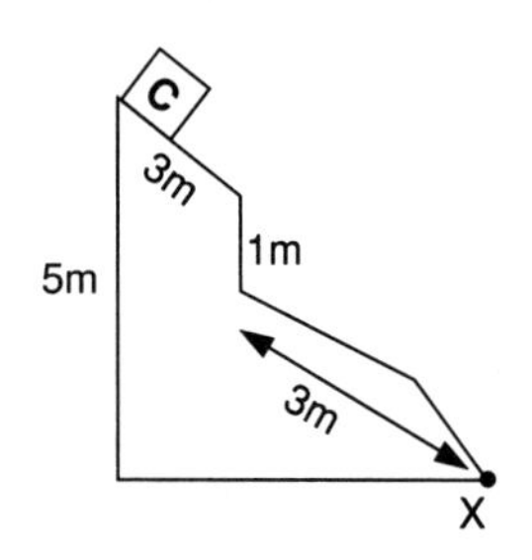

A. A
B. B
C. C
D. All the same speed

26. A glacier floats in a lake such that ¾ of its mass is below the surface of the water. What happens to the level of water in the lake as the glacier melts?

A. It increases.
B. It decreases.
C. It stays the same.
D. None of the above

27. A car slams on its brakes while traveling down a hill with angle X. What is the coefficient of friction if the car stops accelerating down the hill? Assume g = acceleration of gravity, m = mass of car.

A. mg sin X
B. mg cos X
C. sin X
D. tan X

28. Two 200-kg projectiles move toward each other at 20 cm/sec for a head-on collision. Assuming that the collision is inelastic, what is the total momentum after the collision?

A. 8000 kg cm/s
B. 4000 kg cm/s
C. 500 kg cm/s
D. 0 kg cm/s

29. A rope can support a weight of 85 N, but breaks with weight over this limit. A small weight is hoisted up this rope with acceleration of 2 m/s^2. The mass of the weight is 7 kg. Does the rope break? Assume g = 10 m/s^2.

A. It breaks.
B. It does not break.
C. Not enough information is provided.

30. A plane flying 900 mph at an altitude of 6400 feet drops a payload above a point X on the ground. How far from point X does the payload land? Assume g = 32 ft/s^2.

A. 5.0 miles
B. 7.5 miles
C. 10.0 miles
D. 12.5 miles

SOLUTIONS

Section I: Comprehensive Quiz

1. **B** Knowing that $g = GM/r^2$, set up a ratio between X and Y. Because G is the universal gravitational constant, it cancels when a ratio between X and Y is set up.

   $X/Y = (a/b^2)/(2a)/(2b)^2 = (a/b^2)/(2a/4b^2) = 1/(1/2) = 2.$

2. **C** The centripetal acceleration, $a_c = r\omega^2$. Plugging values into this equation gives:

   $a_c = (2m)(5\ rad/s)^2 = 50\ rad^2m/s^2$

3. **A** Recall that Pascal's principle says that a pressure applied to an enclosed static liquid is transmitted undiminished throughout the container. Because P = F/A, write the equation: $F_A/A_A = F_B/A_B$. Note that a large surface area at piston A maximizes the resulting force F produced at A.

4. **D** Power (P) = $\Delta W/\Delta t$. Solve this problem in several steps. First determine the work performed. Note that PE is being converted to KE and heat (work) as the block slides down the ramp. Using the principle of energy conservation, the equation is: $mgh = \frac{1}{2}mv^2 + W$; therefore, $W = mgh - \frac{1}{2}mv^2$. Plugging in values from the problem yields:

   $W = (1\ kg)(10\ m/s^2)(2\ m) - \frac{1}{2}(1\ kg)(4\ m/s)^2 = 12\ J$

   Power = $\Delta W/\Delta t$ = 12 J/10s = 1.2 W

5. **A** This statics problem is best solved using a vector diagram. Remember to set the sum of the x-forces equal to zero and the sum of the y-forces equal to zero.

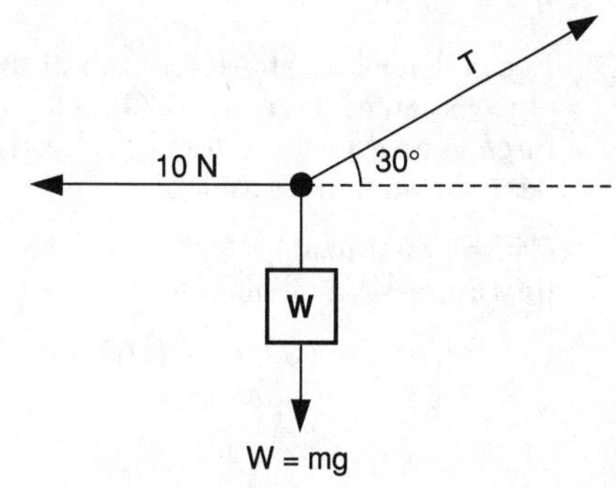

$\Sigma Fx = T\cos 30° - 10\ N = 0 \quad T = 11.5\ N$

$\Sigma Fx = T\sin 30° - mg = 0 \quad W = 5.75\ N$

6. **D** This problem is difficult. Notice that the 200 N force is applied 37° above the horizontal. Therefore, the force has both horizontal and vertical components. The vertical component acts to pull upward on the block, effectively "lightening" or reducing the amount of weight in contact with the floor. This upward pull component, which acts to reduce the final weight of the object in contact with the surface, must be considered (i.e., subtracted. Look at the following vector diagram:

   Because F = ma:

   $200\cos 37° - \mu(mg - 200\sin 37°) = ma$

   $160 - 0.3(180) = ma$

   $106 = 30a \quad a = 3.53\ m/s^2$

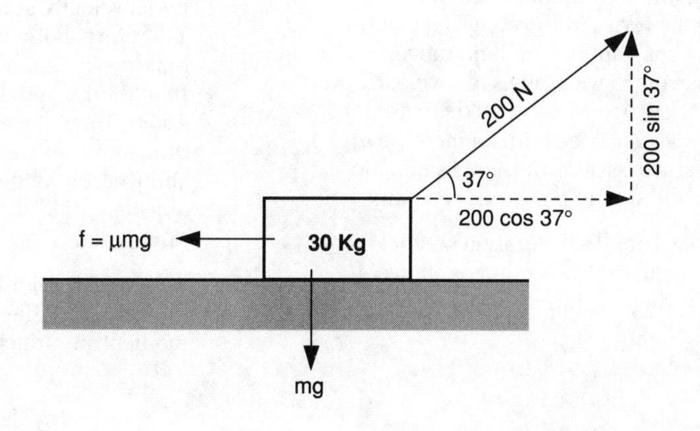

7. **A** In static situations, for every action force there is an equal and oppositely directed reaction force (Newton's third law). In this situation, the table must exert a reaction force of (15 kg)(10 m/s^2) = 150 N.

8. **A** Remember to ignore the work done in accelerating the box from rest to walking speed. Therefore, the work performed in lifting the box = W = F$s\cos\theta$ = (10 kg)(10 m/s^2)(2.5 m)(cos 0°) = 250 J. Note that when the man walks, the force is directed upward whereas the mass moves horizontally (θ = 90°; therefore, cos 90° = 0); in this situation, there is no contribution to work.

9. **A** The tension (T) in the string is T = f = μN = 10 Mg.

10. **C** Recall that all weight (force) acts at the object's center or gravity (CG) as a point force. A coordinate system has already been drawn in the problem.

    CG = {$(\text{dist})(\text{mass})_R$ + $(\text{dist})(\text{mass})_T$}/total mass

    CG = {(50 mm)(100 g) + (⁻60 mm)(70 g)}/170 g = ⁺4.7 mm.

    Therefore, the CG is 4.7 mm to the right of the origin.

11. **A** The kinematic equation needed to solve the problem is: $y - y_o = v_o t + \frac{1}{2}at^2$.

    Because the object is shot from and lands on the ground, $\Delta y = 0$.

12. **A** Tensile strain = $\Delta L/L_o$ = (0.5 m)/(5 m) = 0.1.

13. **A** Because v_o = 6 m/s, a = 2 m/s^2, and t = 3 sec, $x - x_o = v_o t + \frac{1}{2}at^2$ = (6 m/s)(3 sec) + ½(2 m/s^2)(3 sec)^2 = 27 m.

14. **B** According to Bernoulli's equation, as fluid moves from narrow to wide pipes, the pressure increases. Recall that the sum of the terms in Bernoulli's equation equals a constant. When fluid moves from narrow to wide pipes, the velocity of fluid flow decreases. This decrease in velocity is associated with an increase in the pressure term of Bernoulli's equation to keep the sum of the terms constant.

15. **C** The velocity and acceleration vectors are perpendicular to each other as shown in the following diagram:

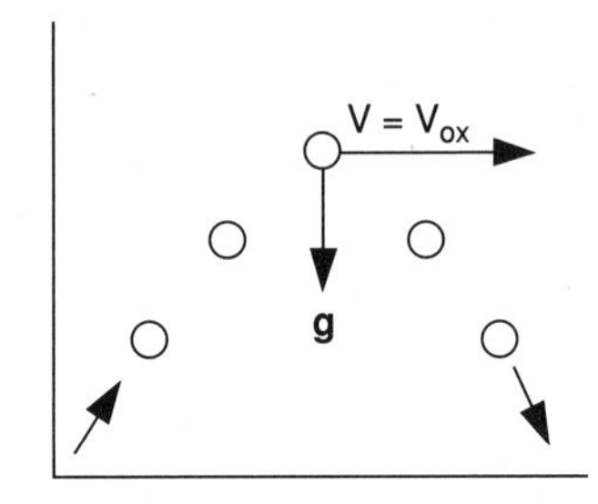

16. **C** The question asks for comparison of the vertical drop during the 2nd second to that during the 1st second. Using one of the kinematic equations, the result is:

    after 1 sec: $y - y_o = v_o t + \frac{1}{2}at^2$

    $= {}^{-}\frac{1}{2}(10)(1)^2 = {}^{-}5$ m

    after 2 sec:

    $= {}^{-}\frac{1}{2}(10)(2)^2 = {}^{-}20$ m

    Therefore, the object drops 5 m during the 1st second and 15 m during the 2nd second. Note that it was necessary to subtract the distance traveled during the 1st second from the distance traveled during the entire 2 seconds to obtain the distance traveled during the 2nd second. The ratio is then 15 m : 5 m = 3 : 1.

17. **D** Specific gravity = $\text{density}_x/\text{density}_{water}$ = (m/v)/(1 gm/cm^3) = [(16 g)/(2 cm^3)]/(1 gm/cm^3) = 2.

18. **C** Although a vacuum contains no air and no air resistance, gravity is still operational. In this question, the acceleration experienced during the flight is g.

19. **B** Recall that gravity affects all objects equally regardless of initial velocity Also note that the objects in this question are given a horizontal velocity. Recall that the x-component of a velocity has no effect on the vertical (y-component) aspect of velocity. In this question, the initial velocity in the y-direction is zero. Therefore, both objects fall from the building as if they are falling from rest. In addition, both balls stay in the air exactly the same amount of time. The time in the air depends only on y-velocity and y-acceleration. Therefore, the objects should strike the ground at the same time.

20. **D** Let the subscript for the man be (m) and the ball (b). Using the conservation of momentum principle:

$m_m v_m = m_b v_b$; then, $v_m = (\frac{1}{4}M)(v)/(12 M) = \frac{1}{48} v$.

21. **C** Constant velocity means zero acceleration. When velocity is zero, acceleration is zero. Both choices A and B are true; therefore, choice C is true.

22. **A** In the first section of the diagram, the displacement is increasing at an exponential rate. This rate of increase implies acceleration, as shown by a positive upward slope on a velocity versus time plot. The midportion of the graph shows displacement increasing at a constant linear rate. This rate of increase implies constant velocity. The last portion of the plot shows a curvilinear change most consistent with a deceleration. This change is shown as a downward line on a velocity versus time plot. These changes are most consistent with choice A.

23. **D** Choices A–C are incorrect statements about speed. Only choice D is correct.

24. **A** Recall the "head-to-tail" technique for adding vectors. Place the tail of vector B to the head of vector A. Draw the addition product from the tail of vector A to the head of vector B. The following diagram shows the solution to this problem.

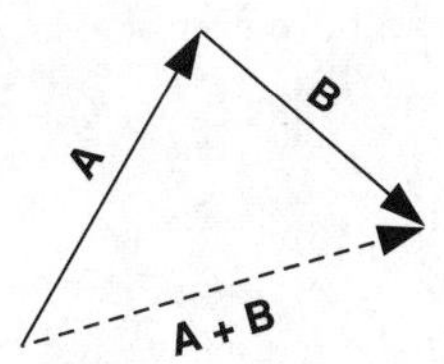

25. **D** Recall that $v = (2\,gh)^{1/2}$ for all free-fall objects and frictionless ramps. Velocity is independent of pathway taken and is solely determined by g and the height above the surface (h).

26. **C** Archimedes' principle of buoyancy says that floating masses (like ice) naturally displace a volume of liquid equal to their own weight. It therefore follows that a full glass of iced tea does not overflow on a hot day.

27. **D** To solve this problem, consider the forces acting on the car. The force directed down the ramp is $mg\sin\theta$. The opposing friction force directed up the ramp is μN or $\mu(mg\cos\theta)$. Because the net acceleration is zero, set the force down the ramp equal to the friction force that opposes it: $mg\sin\theta = \mu(mg\cos\theta)$. Rearranging, $\mu = \tan\theta$.

28. **D** Because these projectiles are of equal mass and are traveling toward each other, the total momentum is: $(200)(20) + (200)(^-20) = 0$.

29. **B** The rope can support only 85 N or less of force (tension). Calculate whether the total force experienced due to the weight of the object alone and the acceleration ($2\ m/s^2$) experienced is greater than 85 N. Following are the two ways to attack this problem:

1. $F = m(a + g) = 7(10 + 2) = 84\ N$
2. $F_{mass} = (7)(10) = 70\ N$
   $F_{accel} = (7)(2) = 14\ N$. The total force $= 70\ N + 14\ N = 84\ N$.

Therefore, the total force or tension is 84 N by either method, and the rope does not break.

30. **A** In this involved two-step problem, first find the time required for the payload to drop 6400 feet. Second, knowing the time spent in the air, calculate the horizontal distance traveled. Beware of units in this problem.

The knowns are: $a_x = 0$. $v_{oy} = 0$. $v_x = v_{ox} = 900$ mph. $y = {}^-6400$ ft.

1. $y = v_o t + \frac{1}{2}at^2$

   $^-6400 = {}^-(\frac{1}{2})(32)t^2$

   $t = 20\ \text{sec} = \frac{1}{3}\ \text{min} = \frac{1}{180}\ \text{hr}$

2. $x = v_{ox}t$

   $x = (900\ \text{miles/hr})(\frac{1}{180}\ \text{hr}) = 5\ \text{miles}$

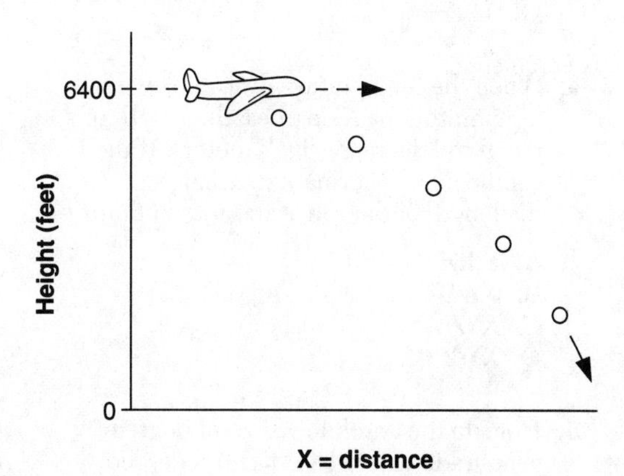

HIGH-YIELD REVIEW QUESTIONS

Section II: Physics

Temperature, Heat, Thermodynamics, and Electrostatics

1. Suppose that a solid, uniform cube of edge 2 meters is heated such that the top of the cube is three times the surface temperature of the base of the cube. What is the ratio of the radiation power emitted from the top to the radiation power emitted from the base?

A. 27
B. $\frac{1}{9}$
C. 81
D. $\frac{1}{27}$

2. 50 grams of ice at 0°C is added to 100 grams of water at 100°C. What is the final temperature of the system, if there is no gain or loss of heat from the environment?

A. $50 > T > 45$
B. $T < 45$
C. $T > 50$
D. Not enough information to be determined

3. If the surface temperature of a star is dropped by a factor of 2, the radiant energy reaching the earth is reduced by a factor of:

A. 8.
B. 4.
C. 16.
D. 32.

4. When the temperature drops below room temperature by X degrees, the length of a metal rod decreases by Y meters. If the coefficient of thermal expansion = Z, what is the length of the rod at room temperature?

A. Y/XZ
B. XZ/Y
C. XYZ
D. XY/Z

5. How do the absolute values of degree centigrade (C), kelvin (K), and degree Fahrenheit (F) compare?

A. C > K > F
B. C = K > F
C. F > C = K
D. F = C > K

6. Two objects, A and B, are dropped into two adiabatic insulators that contain equal volumes of liquid X at 100°C. Object B has a higher heat capacity than object A. At equilibrium, which object has a higher temperature?

A. A
B. B
C. Both the same temperature
D. None of the above

7. Five grams of H_2O steam is added to 5 grams of H_2O ice. The heat required to melt the ice:

A. is greater than the heat required to condense the steam.
B. is less than the heat required to condense the steam.
C. is equivalent to the heat required to condense the steam.
D. cannot be compared to the heat required to condense the steam.

8. During an isothermal process:

A. $\Delta W = 0$.
B. $\Delta G = 0$.
C. $\Delta U = 0$.
D. $\Delta S = 0$.

9. Heat energy is supplied to the gas in the following piston system. The atmosphere then exerts a force at point A that compresses the gas. The change in internal energy is:

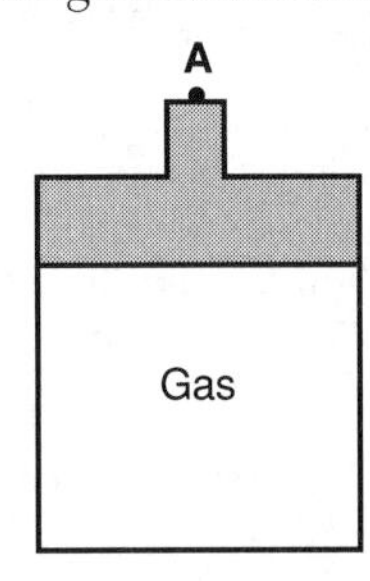

A. negative.
B. positive.
C. zero.

10. Given the following *pressure–volume curve* for gas X, which is the closest approximate value for the work done in the adiabatic expansion (from A to B) of gas X?

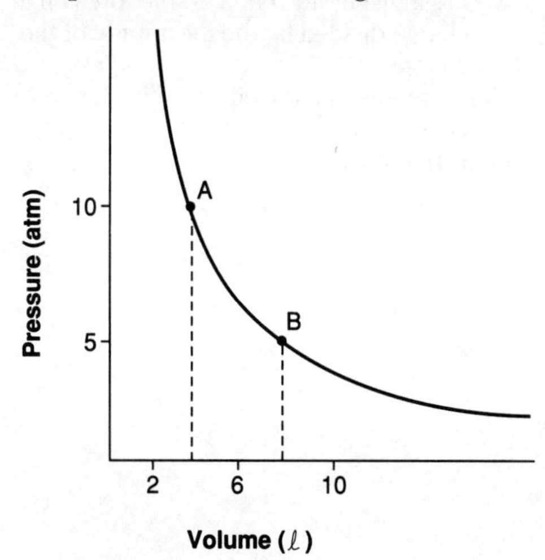

A. 20 l atm
B. 20 J
C. 30 l atm
D. 50 J

11. The following graph is the PV curve for a Carnot engine. Which sections represent adiabatic processes?

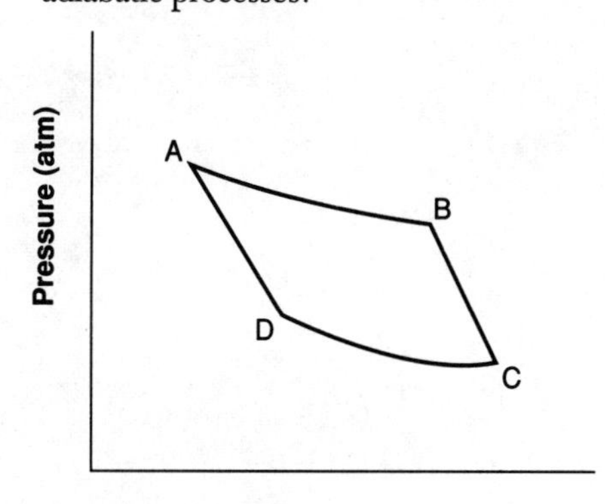

A. AB, CD
B. AB, BC
C. BC, AD
D. AD, AB

12. In all isochoric processes, which equation is true?

A. $\Delta U = 0$
B. $\Delta S = 0$
C. $W = 0$
D. $P = 0$

13. When a liquid X is frozen, its entropy:

A. increases.
B. decreases.
C. stays the same.
D. cannot be determined.

14. A point charge "P" experiences a force of 10 N to the right when a point charge of $^{+}5\ \mu C$ is placed 2 meters to the right of "P". What is the point charge "P"? Assume $k = 9 \times 10^9$ Nm^2/C^2.

A. $\frac{8}{9} \times 10^{-3}$ C
B. $\frac{4}{5} \times 10^{-3}$ C
C. $\frac{1}{3} \times 10^{-3}$ C
D. $\frac{7}{12} \times 10^{-3}$ C

15. Which statement accurately describes conductors?

A. They have no free electrons.
B. They have relatively few electrons whose number depends on specimen temperature.
C. They exert electrostatic forces on one another.
D. They have some electrons that are relatively mobile.

16. Which statement accurately describes electric dipoles?

A. They are found in electrically neutral objects.
B. They are vectors.
C. They are diverted from negative to positive centers of charge.
D. All of the above are correct statements.

17. Three equal positive point charges, A, B, and C, are placed on three corners of a square. What is the ratio of the forces AB to AC ($F_{AB}:F_{AC}$)?

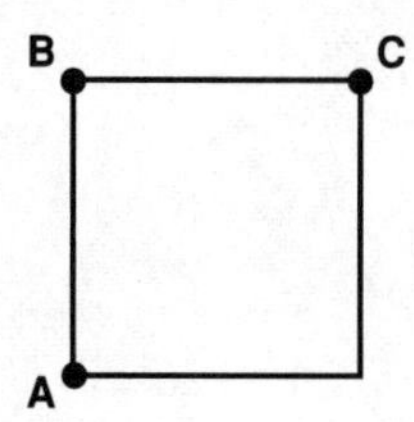

A. 1:1
B. $\sqrt{2}$
C. 2:1
D. $\sqrt{2/2}$

18. Point charges X, Y, and Z, all of the same charge and magnitude, sit on an equilateral triangle. The net force on point Y is:

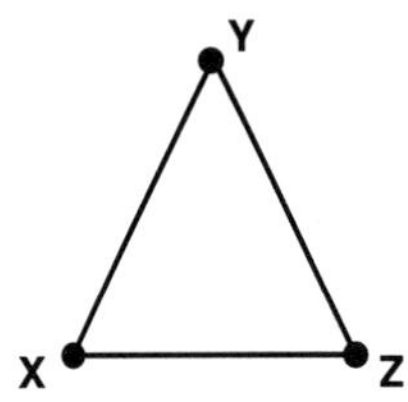

A. continuous with XY.
B. continuous with YZ.
C. parallel to XZ.
D. perpendicular to XZ.

19. A negative charge would experience no net force if placed at:

B ●

A. A.
B. B.
C. C.
D. none of the above.

20. An electric field:

A. is the force acting on a negative test charge.
B. has a magnitude equal to the force on a charge divided by the magnitude of the charge.
C. is the force acting on a positive test charge.
D. is B and C.

SOLUTIONS

Temperature, Heat, Thermodynamics, and Electrostatics

1. **C** The power emitted = σEAT^4.

   Because the area is equivalent on top and bottom:

   Cube top: $\sigma EAT^4 = 3^4 = 81$

   Cube bottom: $\sigma EAT^4 = 1^4 = 1$

   Therefore, the ratio of the radiation power emitted from the top compared to the base is 81:1.

2. **B** Solve this problem by setting the heat gain equal to the heat loss.

   heat gain = heat loss

   ice melt + water warm = water cool

   $(50g)(80\ cal/g) + (50\ g)(1\ cal/g°C)(T_f - 0) = (100\ g)(1\ cal/g°C)(100 - T_f)$

   $4000\ cal + 50\ T_f = 10{,}000 - 100\ T_f$

   $150\ T_f = 6000$

   $T_f = 40°C$; therefore, $T_f < 45°C$

3. **C** Radiant energy = σEAT^4. If the temperature drops by a factor of 2 or by half, the only factor that changes is T. The radiant energy relationship shows the temperature term is taken to the power of four. Therefore, $T^4 = (½)^4 = \frac{1}{16}$. This ratio corresponds to a reduction by a factor of 16.

4. **A** $\alpha = (1/L_o)(\Delta L/\Delta T)$. Thus, $Z = (1/L_o)(Y/X)$. Solving for Y gives: $(L_o)(XZ) = Y$. Solving for the length of the rod at room temperature (L_o) gives: $L_o = Y/XZ$.

5. **B** 1°C change = 1K change. Both a 1°C and 1K change are greater than a 1°F change. This relationship is apparent because 100°C separate ice and boiling water, whereas 182°F (32° –> 212°F) separate ice and boiling water. Therefore, each °F is less than a °C.

6. **A** Objects with a higher heat capacity store more energy (calories) per increase in temperature. Therefore, A is hotter than B.

7. **B** It requires 540 cal/g to convert liquid H_2O to steam, and 540 cal/g is released in condensing steam. Only 80 cal/g is needed to melt ice. Steam condenses at 100°C and ice melts at 0°C.

8. **C** $\Delta U = 0$ during an isothermal process because U, the internal energy of a substance, depends only on temperature. Work = 0 if $P\Delta V = 0$, because $W = P\Delta V$. $\Delta G = 0$ for processes in equilibrium. $\Delta S = 0$ if there is no change in entropy, or no change in the randomness of a system. Randomness can change even if temperature is constant (e.g., volume changes).

9. **B** The first law of thermodynamics says that $\Delta U = \Delta Q - \Delta W$. Recall that ΔW is the work done by the system (positive work for expansions), and ΔQ is energy supplied to the system as heat. Because heat is supplied to the system, ΔQ is positive. Furthermore, the system is acted on by external compression forces, so ΔW is negative (work done on system). Therefore, $\Delta U = (+) - (-) =$ a positive quantity.

10. **C** The work given on a PV curve equals the area under the curve between the points of interest:

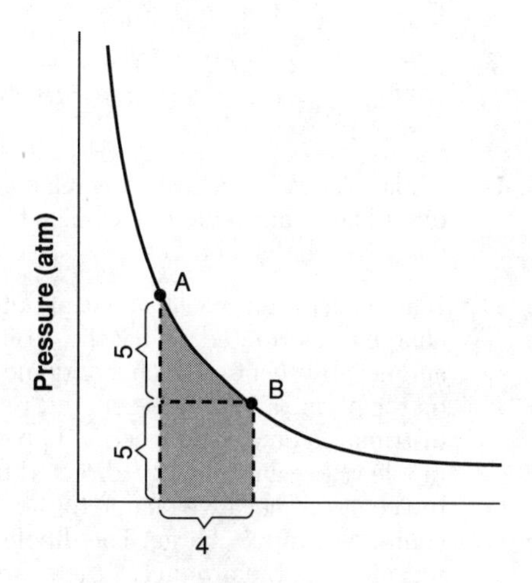

   W = (5 atm)(4 liters) + (½)(5 atm)(4 liters) = 30 l atm. Joules are not units here (1 J = 1 Nm), because the problem provides units in the British system.

11. **C** Carnot described an engine that progresses through four PV changes before returning to its original state. Starting at A, an isothermal expansion occurs as the gas in the engine comes in contact with a heat reservoir. The gas does not heat up because it is allowed to expand. At B, the gas expands adiabatically, neither gaining nor losing heat. The temperature decreases because the gas expands. At C, the gas is isothermally compressed as it gives up heat to the heat reservoir. Finally, the gas is compressed adiabatically from D to A. The overall work done by the engine in one cycle equals the area bounded by the shaded figure ABCD.

12. **C** Isochoric means constant volume. Because $W = P\Delta V$, if $\Delta V = 0$, $W = 0$.

13. **B** Entropy means randomness of the system. Because freezing a liquid allows less random motion of molecules than the liquid phase, entropy decreases.

14. **A** Set up a diagram for the situation described:

•————————> •
P (force) $^{+}5\ \mu C$

Because P is attracted toward the $^{+}5\ \mu C$ charge (force is to the right), P has a negative charge. The force between two point charges in air is given by Coulomb's law:

$F = kq_1q_2/r^2$; therefore, $q_1 = Fr^2/kq_2$.

$P = (10\ N)(2\ m)^2/(9 \times 10^9\ Nm^2/C^2)(5 \times 10^{-6}\ C) = (40\ Nm^2)/(45 \times 10^3\ Nm^2C^{-1}) = 8/9 \times 10^{-3}\ C$.

15. **D** Choice A describes insulators. Choice B describes semiconductors. Choice C describes charged bodies.

16. **D** If an object is electrically neutral but the charge is distributed so that the positive and negative centers of charge do not lie on top of one another, the charge distribution creates an electric dipole. A dipole is a vector quantity, directed from the center of negative charge to the center of positive charge. The dipole magnitude is the product of the charges and the distance separating the centers of charge.

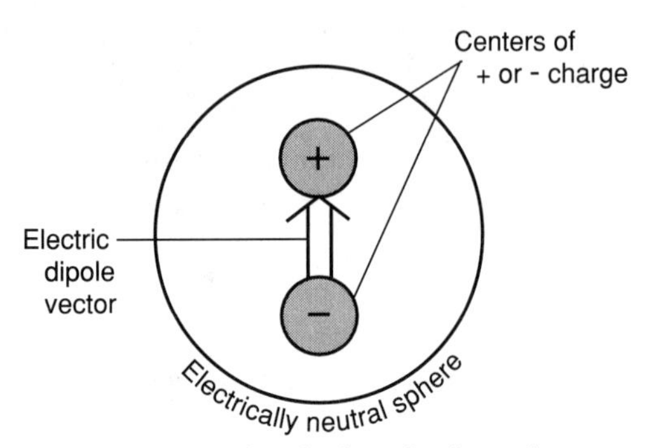

17. **C** Suppose that the length of an edge is 1 m.

The force between A and B is $F = kq_1q_2/(1)^2 = kq^2$.

The force between A and C is $F = kq_1q_2/(\sqrt{2})^2 = kq^2/2$.

The ratio then is: $(kq^2)/(kq^2/2) = 2$.

18. **D** The following diagram shows the resultant vector, which is perpendicular to XZ.

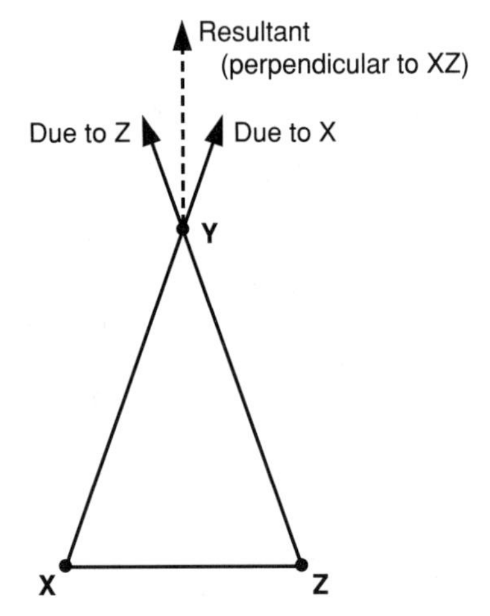

19. **D** Point A would have horizontal force vectors cancel, but would have net movement vertically:

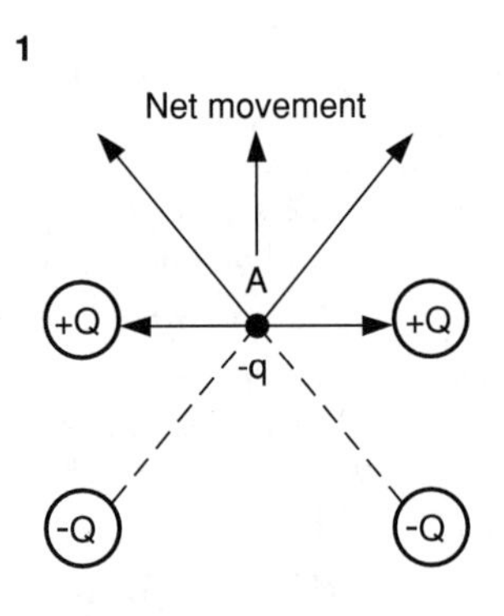

Similary at B, repulsion from ($^-$Q) charges gives vertical displacement:

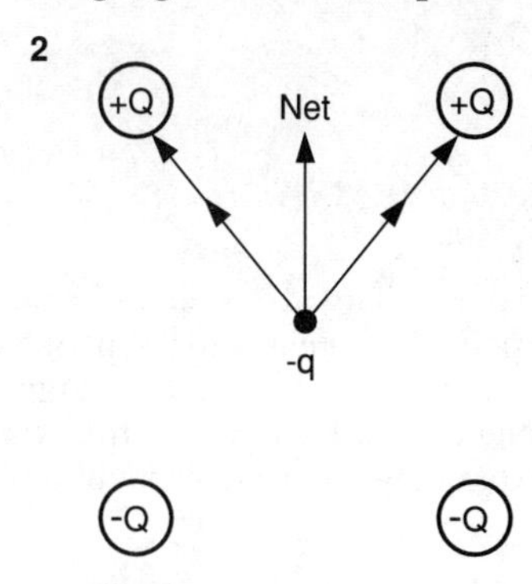

Similary, at point C, no horizontal movement occurs, only vertical movement:

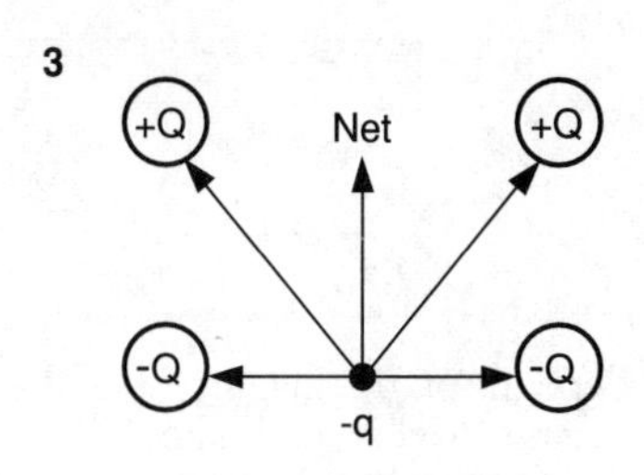

20. **D** The electric field is defined as the force that a positive test charge experiences at a point in space. The electric field points in the direction that a positive charge is pushed if placed at a point in space. Electric field lines flow from positive charges to negative charges. If ($^+$q) is a positive charge being acted on by a force F caused by charge (Q):

$E = F/q = KQ/r^2$

where K is 9×10^9 Nm^2/C^2 and r is the distance between the test charge and (Q). If Q is negative, E points toward ($^-$Q).

HIGH-YIELD REVIEW QUESTIONS

Section II: Physics

Electrostatics, Electric Fields, Capacitance, and DC Circuits

1. What are the units of the volt?

A. N/C
B. N/W
C. W/C
D. J/C

2. A uniform electric field is applied to a charge of $^{-}5 \times 10^{-4}$ C. A 5×10^{-2} N force must be applied to this charge to keep it at rest. What is the magnitude of this electric field?

A. 100 V/m
B. $^{-}1 \times 10^{-2}$ C/N
C. 1×10^{-2} N/C
D. $^{-}25 \times 10^{-8}$ N C

3. Four equal charges of 1×10^{-9} C are placed on the four corners of a square of edge length $\sqrt{2}$ meters. What is the potential at the center of the square? Assume $k = 9 \times 10^9$ Nm^2/C^2.

A. 9 V
B. 9×10^{-9} V
C. 36 V
D. 3.6×10^{-8} V

4. At a point in space:

A. if the electric field > 0, the potential must be > 0.
B. if the potential > 0, the electric field must be > 0.
C. if the electric field > 0, the potential may be 0.
D. if the potential < 0, the electric field may be < 0.

5. If the electric field in a region is 0, the potential is:

A. 0.
B. constant.
C. proportional to r.
D. proportional to q.

6. Which one is NOT a unit of energy?

A. Nm
B. Cal
C. Ws
D. V

7. A positively charged test charge is moved from point A to point B in the path indicated. Two charges of $^{-}5$ C and two charges of $^{-}2$ C sit along the path equidistant from A and B. The work done in this movement from A to B:

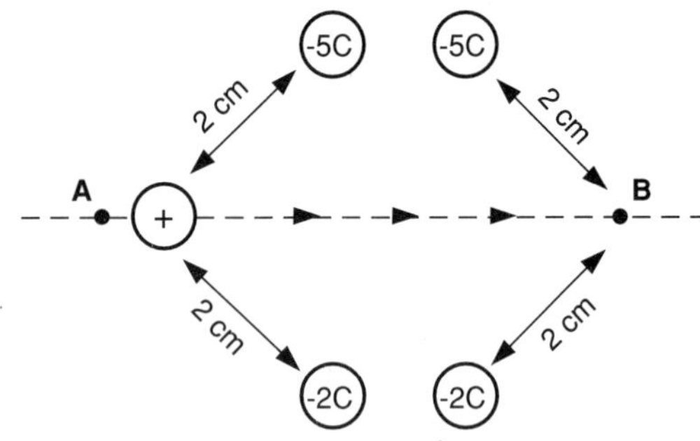

A. depends on the magnitude of the negatively charged bodies.
B. depends on the charge of our test charge.
C. depends on the distance between negative charges.
D. is zero.

8. The work done on moving a charged body through an electric field between two points depends on:

A. the magnitude of the charge on the body.
B. the potential difference between the points.
C. the distance between the points.
D. A and B.

9. A capacitor:

A. is two insulating surfaces separated by a conductor.
B. is two conducting surfaces separated by an insulator.
C. does not maintain charge separation.
D. has a capacitance that relates surface area to voltage.

10. Three capacitors are connected in parallel. The equivalent capacitance is:

A. less than if the capacitors were connected in series.
B. equivalent to the average capacitance of the three capacitors.
C. greater than the capacitance of any of the three single capacitors.
D. none of the above.

11. Suppose that a dielectric slab (k = 4) is inserted between the plates of a parallel-plate air capacitor. The stored energy on the plates:

A. remains the same.
B. decreases by a factor of 4.
C. increases by a factor of 4.
D. increases by a factor of 2.

12. Suppose that a parallel-plate capacitor system is fully charged. As the plates are separated, one would expect:

A. the electric field and charge on the plates to decrease.
B. only the electric field to decrease.
C. the electric field to stay constant, but the charge on the plates to decrease.
D. the electric field to decrease, but the charge on the plates to stay constant.

13. Which capacitor system stores the most energy?

Assume that the voltage across X and Y is identical and that individual capacitors are identical in each system.

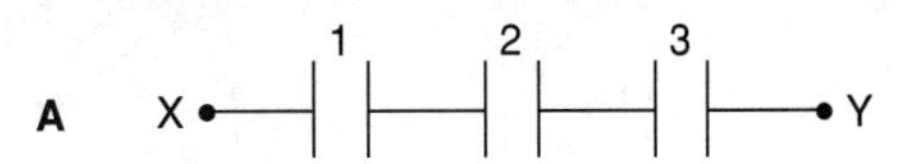

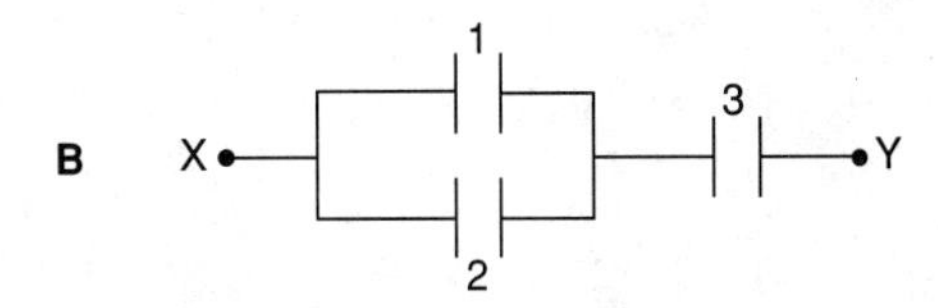

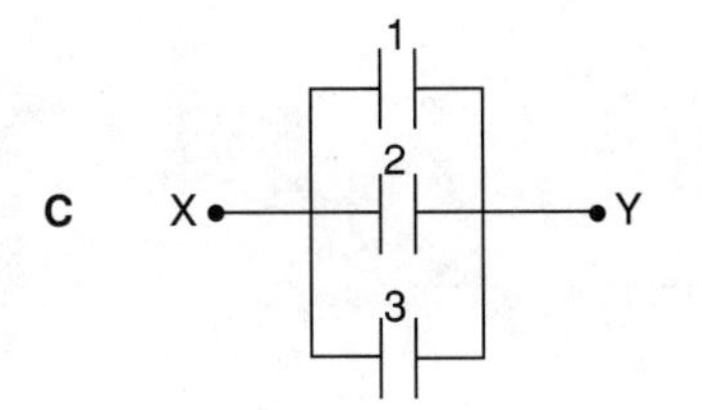

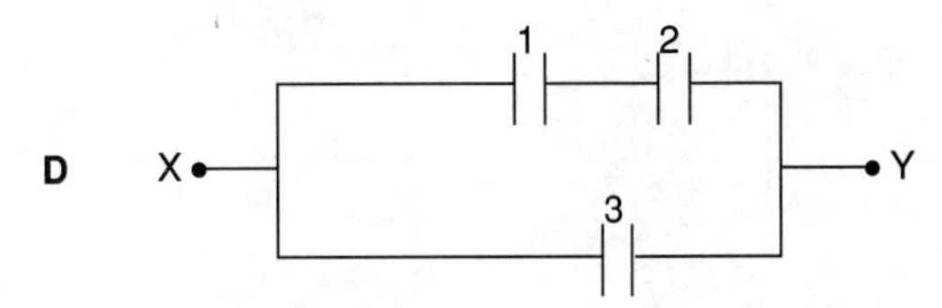

14. Two wires of identical composition and length have different resistances because of differing radii. If the two wires have resistances of 16 Ω and 4 Ω, respectively, what is the ratio of their radii?

A. $\frac{1}{16}$
B. $\frac{1}{8}$
C. $\frac{1}{4}$
D. $\frac{1}{2}$

15. Circuits A and B are both powered by a 5 V battery. Individual resistors are all 1 Ω. What is the ratio of the current flowing through R_1 to that flowing through R_3?

A

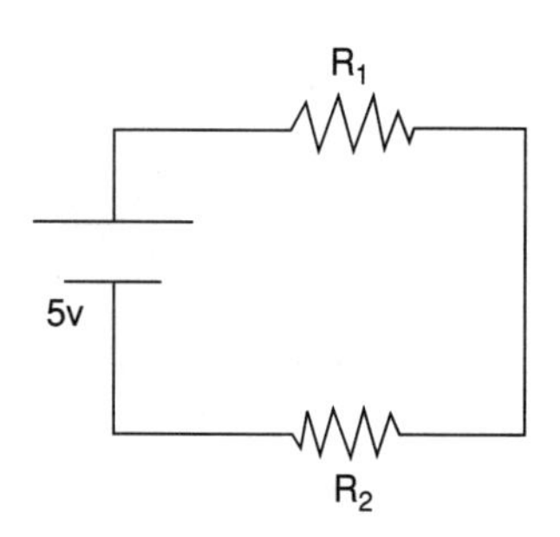

B

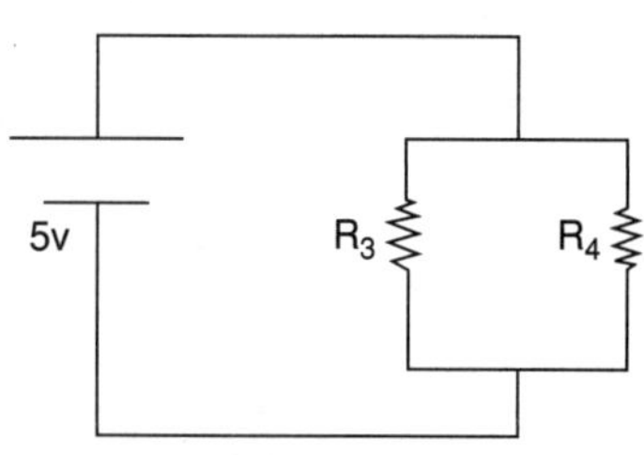

A. $\frac{1}{3}$
B. $\frac{2}{1}$
C. $\frac{1}{2}$
D. $\frac{3}{1}$

16. A series circuit is connected to a 10 V EMF source. If the 2 Ω resistor is replaced with a 4 Ω resistor, what voltage change occurs from A to B because of this replacement?

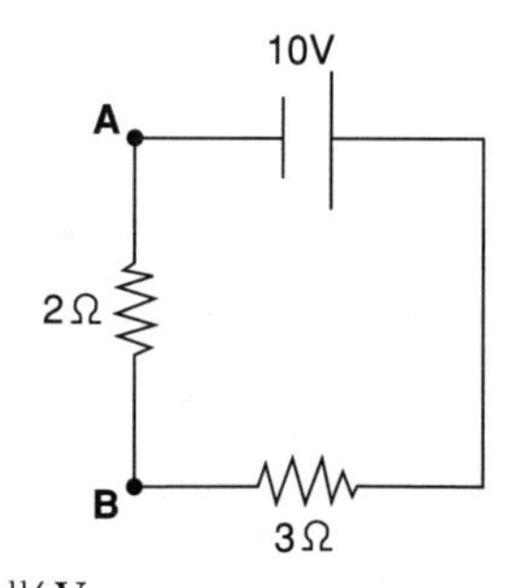

A. $\frac{11}{7}$ V
B. $\frac{12}{7}$ V
C. $\frac{13}{7}$ V
D. 2 V

17. Suppose that the current associated with this circuit flows through the 2 Ω resistor. If a 4 Ω resistor is added at A, the current flowing through the 2 Ω resistor changes by how many amps?

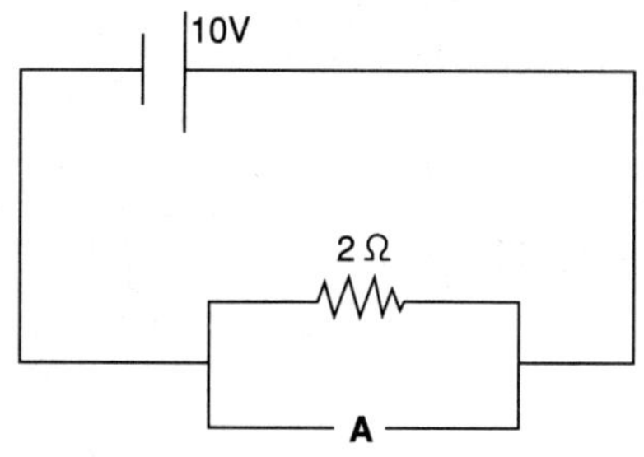

A. 4 A
B. 2.5 A
C. 2 A
D. 0 A

18. Given the following diagrammed circuit, what is the circuit current?

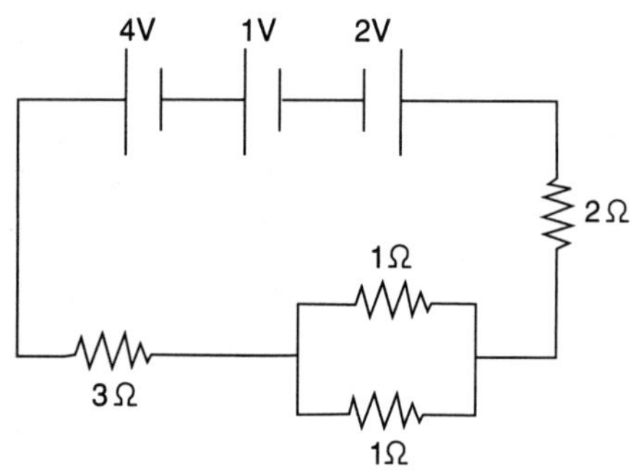

A. $\frac{5}{13}$ A
B. $\frac{3}{4}$ A
C. $\frac{5}{7}$ A
D. $\frac{6}{11}$ A

19. If a 2-amp current flows through the 6 Ω resistor, what is the EMF of the battery if it has no internal resistance?

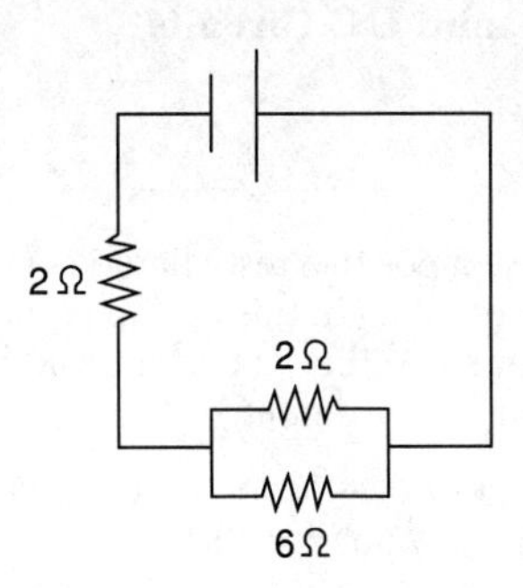

A. 20 V
B. 10 V
C. 5 V
D. None of the above

20. If 8 amps flow through the 2 Ω resistor, what current flows through the 4 Ω resistor?

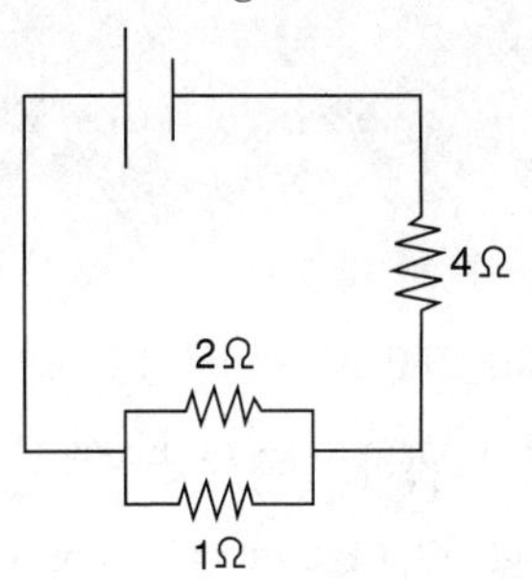

A. 16 A
B. 24 A
C. 36 A
D. 48 A

SOLUTIONS

Electrostatics, Electric Fields, Capacitance, and DC Circuits

1. **D** Recall that 1 volt = 1 Nm/C = J/C.

2. **A** The magnitude of this electric field is found by using $E = F/q$. Plugging in the values of F and q given in the question: $(5 \times 10^{-2}\ N)/(5 \times 10^{-4}\ C) = 100\ V/m$. (Recall that 1 V = 1 Nm/C).

3. **C** The potential due to a single charge is KQ/r. Based on the diagram of the following problem, r = 1 meter.

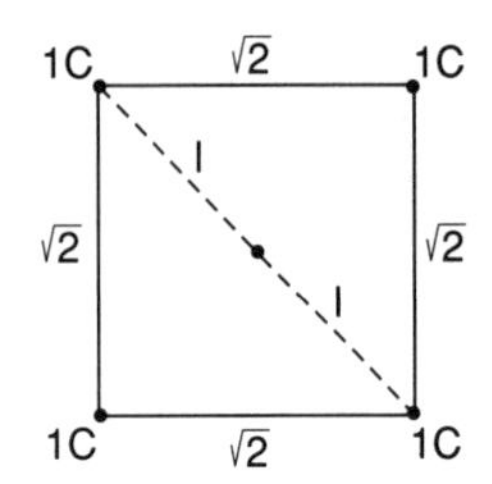

Also note that the potential due to four charges is four times the potential due to one charge. To calculate the solution to this question, find the potential due to a single given charge, and multiply that value by 4.

$(9 \times 10^9\ Nm^2/C^2)(1 \times 10^{-9}\ C)/1\ m = (9\ V)(4\ charges) = 36\ V$.

4. **C** For this question draw a diagram with charges and determine which choice must be true. A diagram with charges follows:

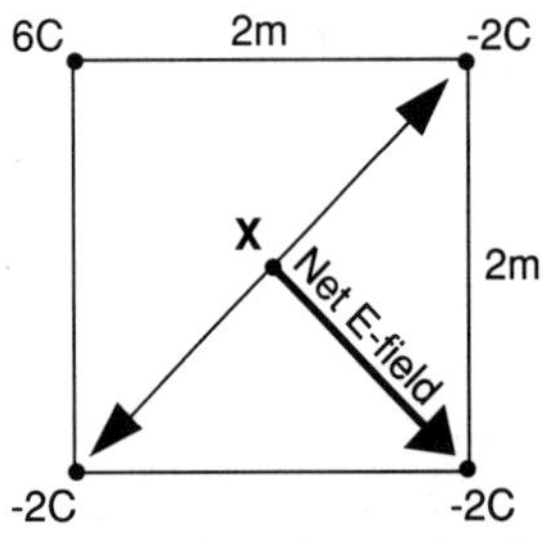

Because potential = KQ/r, for the given charges, the potential would be:

$\{K(6\ C + (^-2\ C) + (^-2\ C) + (^-2\ C)\}/r = 0$.

Therefore, at point X, the potential is zero. The diagram shows the direction of the E-field. Calculate the magnitude of the E-field by noting which charges cause a positive test charge net movement. (In this example, the 6 C charge and the $^-2$ C charge across from it cause movement.)

$E = KQ/r^2 + KQ/r^2 = K(2\ C)/(\sqrt{2}\ m)^2 + K(6\ C)/(\sqrt{2}\ m)^2 = 8K/2 = 4K$.

Because $K = 9 \times 10^9\ Nm^2/C^2$, $E = 4(9 \times 10^9\ Nm^2/C^2) = 3.6 \times 10^{10}\ V/m$.

Note that the E-field is greater than zero. Therefore, choice C is true. The other choices are incorrect. One can draw a diagram with different charges and rule out the other choices.

5. **B** A potential is defined as the negative of the work done by an electric field in moving a positive unit charge a distance. Since no E-field is present, the potential at any one point is constant compared to any other point. If the question had asked for the potential difference between two points, the answer would have been zero.

6. **D** Recall that 1 Nm = 1 J. Joules and calories are units of energy. Therefore, choices A and B are units of energy. 1 W = 1 J/s; therefore, 1 Ws = 1 J/s(s) = J. This equation shows that choice C is a unit of energy. Choice D is not a unit of energy because a volt = J/C, which is a measure of electrical potential.

7. **C** The work done in moving from A to B equals work (W), where $W = q\Delta V$, q is the charge magnitude of the test charge, and ΔV is the potential difference between points A and B. Because the potential difference depends on the charges setting up an electric field, and the net charges acting on A and B are identical (symmetric geometry), there is no potential difference from A to B. Therefore, $\Delta V = 0$, and $q\Delta V = W = 0$.

8. **D** Work = $q\Delta V$, where q is the charge of the body, and ΔV is the potential difference between the two points of travel. Statements given in choices A and B are both correct, making choice D the best choice.

9. **B** This definition is the one for a capacitor. Capacitors often consist of two parallel plates separated by a small distance and insulated by poor conductors such as air. The magnitude of the E-field between the plates = E = V/d, where V is the potential difference between the plates (volts), and d is the distance between the plates. C (capacitance) = Q/V, where Q is the charge on one plate. The farad (F) = 1 C/V and is the unit of capacitance. C = Q/V = $K\epsilon_o$ A/d where K = dielectric constant, ϵ_o = constant, A = area of plate, and d = distance between plates. The dielectric constant (K) describes the insulating properties of various substances that can be placed between the plates. $K_{air} = 1.0$, $K_{water} = 78$.

For reference, the potential energy stored on a capacitor = ½QV = ½CV^2 = ½Q^2/C.

10. **C** To answer this question, one must understand the relationships for equivalent capacitance:

ΣC in parallel:

$C_{total} = C_1 + C_2 + C_3$

ΣC in series:

$1/C_{total} = 1/C_1 + 1/C_2 + 1/C_3$

Based on these relationships, only choice C is correct.

11. **B** The potential energy on a capacitor = ½QV. Recall that the capacitance increases by a factor of 4 if K = 4 (C = $K\epsilon_o$ A/d). Because V = Q/C, if the capacitance increases by a factor of 4, V decreases by a factor of 4. Because PE = ½QV and Q is constant, PE must decrease by a factor of 4.

12. **A** E = KQ/r^2, where r = distance between the charges. Because the plates pull away from each other, r increases and E decreases. Because Q = CV, and capacitance depends on the distance between the plates (C = $K\epsilon_o$ A/d), capacitance decreases and Q (charge) decreases as distance increases.

13. **C** Suppose that the capacitance of each capacitor = 1.0.

In choice C: $C = C_1 + C_2 + C_3 = 1 + 1 + 1 = 3$.

In choice B: $C_{1,2} = 1/C = 1/1 + 1/1 = 1/2$.
$C_{1,2} + C_3 = 1/2 + 1 = 1\frac{1}{2}$.

In choice A: $C_{1,2,3} = 1/C_1 + 1/C_2 + 1/C_3 = 1/C_{1,2,3} = 1/3$.

In choice D: $C_{1,2} = 1/C_1 + 1/C_2 = 1/2$.
$C_{total} = C_3 + 1/2 = 1 + 1/2 = 3/2$.

Because E = ½ CV^2 and the voltages are equal, choice C has the greatest energy because it has the greatest capacitance.

14. **C** R = l/A. Because l/A is constant, R ≈ l/A ≈ $l/\pi r^2$ ≈ l/r^2.

$$\frac{R_1}{R_2} = \frac{16}{4} = \frac{1/r_1^2}{1/r_2^2} = \frac{16}{r_2^2} = \frac{4}{r_1^2}$$

$$16\,r_1^2 = 4\,r_2^2$$

Setting up a ratio:

$$\frac{r_1^2}{r_2^2} = \frac{1}{4} \Rightarrow \frac{r_1}{r_2} = \frac{1}{2}$$

15. **C** For circuit A, V = IR. 5 V = I(2 Ω).

I = 5/2 A (current through R_1)

For circuit B, V = IR. $R_T = 1R_T = 1/1 + 1/1$. R_T = ½ Ω.

5 V = (½ Ω)(I) I = 10 A (current through the parallel resistors).

Determine the amount of current through R_3; remembering that current splits between elements of parallel circuits in an inverse ratio to the resistances. Because both R_3 and R_4 have equivalent resistances, the total current of 10 A splits equally between them. Therefore, the current flowing through is 5 A.

The ratio between the current flowing between R_1 and R_3 is: (5/2 A)/5 A = ½.

16. **B** Find the original voltage that is associated with the 2 Ω resistor.

R_T = 2Ω + 3Ω = 5Ω. I = V/R = 10 V/5 Ω = 2 A.

Therefore, 2 A flows across the 2 Ω resistor. The potential drop across the 2 Ω resistor = V = IR = (2 A)(2 Ω) = 4 V.

Now, find the new voltage associated with a 4 Ω resistance:

RT = 4 Ω + 3 Ω = 7 Ω

I = V/R = 10 V/7 Ω = 10/7 A.

V = (10/7 A)(4 Ω) = 40/7 V

The voltage change is now: (40/7 V) − 4 V = 12/7 V.

17. D Current before the change: $V = IR$, $I = V/R$. $I = 10\ V/2\ \Omega = 5\ A$.

Current after change: Start by finding the resistance.

$1/R_T = \frac{1}{2}\ \Omega + \frac{1}{4}\ \Omega = 1/R_T = \frac{3}{4}\ \Omega$
$R_T = \frac{4}{3}\ \Omega$

Now, note that current flowing through the parallel resistance system $= I = V/R = 10\ V/(\frac{4}{3}\ \Omega) = 7.5\ A$.

Set up a ratio to find how much of the current flows through the 2 Ω resistor:

Resistance	Resistance ratio	Current ratio	
2 Ω	1	2	$\frac{2}{3}$
4 Ω	2	1	$\frac{1}{3}$

Because total current $I = 7.5\ A$, and the 2 Ω resistor allows twice as much current to flow as the 4 Ω, $I = \frac{2}{3}(7.5\ A) = 5\ A$.

Therefore, the change equal to the current flowing before the change (5 A) minus the current flowing after the addition of the 4 Ω resistor: $5\ A - 5\ A = 0\ A$.

18. D Note that the 4 V and 1 V battery run one way, and the 2 V battery runs the other way. Subtract 2 V from the sum of 4 V and 1 V. $5\ V - 2\ V = 3\ V$ net voltage.

Now find the circuit resistance.

Parallel resistance: $1/R = \frac{1}{1} + \frac{1}{1}$
$R = \frac{1}{2}\ \Omega$

Total resistance = sum of series resistance and net parallel resistance: $3 + \frac{1}{2} + 2 = \frac{11}{2}\ \Omega$

$I = V/R = 3\ V/(\frac{11}{2}\ \Omega) = \frac{6}{11}\ A$

19. D If 2 A flows through the 6 Ω resistor, use a ratio table to find the total current that flowed through the circuit.

Resistance	Resistance ratio	Current ratio	
6 Ω	3	1	2 A
2 Ω	1	3	(2 A)(3) = 6 A.

The total current through the parallel system is therefore: $2\ A + 6\ A = 8\ A$. If 8 A flows through the parallel system, 8 A flows through the total circuit resistance.

The total circuit resistance $= 2\ \Omega +$ parallel resistance.

The parallel resistance is: $1/R = \frac{1}{6}\ \Omega + \frac{1}{2}\ \Omega = \frac{3}{2}\ \Omega$.

$R = 2\ \Omega + \frac{3}{2}\ \Omega = \frac{7}{2}\ \Omega$.

The battery voltage (EMF) is therefore: $V = IR = (8\ A)(\frac{7}{2}\ \Omega) = 28\ V$.

20. B Use a ratio table to find the total current that flowed through the circuit.

Resistance	Resistance ratio	Current ratio	
2 Ω	2	1	($\frac{1}{3}$) 8 A
1 Ω	1	2	($\frac{2}{3}$) 16 A

Therefore, the total circuit current is $8\ A + 16\ A = 24\ A$. Because the same current flows through each member of a series circuit, 24 A flows through both the parallel component of the circuit (it is in series with the 4 Ω resistor) and the 4 Ω resistor.

Section II: Physics

DC Circuits and Electromagnetism

1. How much power is dissipated when a 10 V battery passes current through the following circuit?

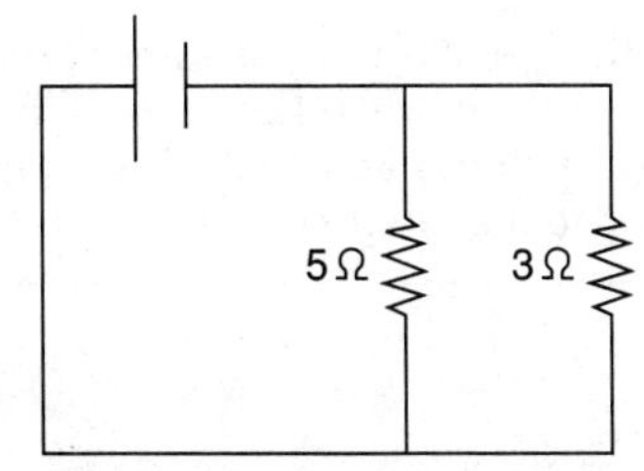

A. $145/2$ W
B. $140/7$ W
C. $152/3$ W
D. $160/3$ W

2. Given a 10-volt EMF and the following circuit, what is the current through the 4 *Ω* resistor?

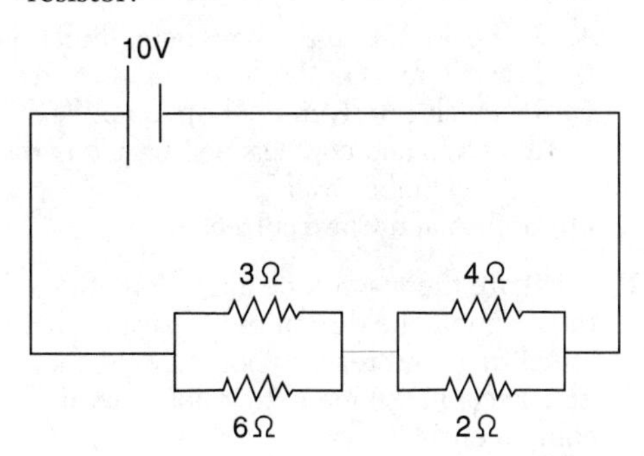

A. 1 A
B. 2 A
C. 3 A
D. 4 A

3. Which statement about magnetism is NOT correct?

A. A magnetic force is produced whenever a charged particle moves within a magnetic field.
B. The direction of the magnetic force is perpendicular to both the velocity and magnetic field.
C. A charged particle forced into circular motion by a magnetic field changes speed.
D. Both A and C are incorrect.

4. A patient known to have left lower lobe pneumonia is given two chest x-rays. The first x-ray, with a wavelength of 0.03 nm, is aimed to give a posterior–anterior (PA) view of the lung fields, whereas the second x-ray, with a frequency of 10^{18} hertz, is aimed to give a left lateral view. Which x-ray possesses more energy?

A. The PA view x-ray
B. The left lateral view x-ray
C. The PA and left lateral view x-rays have the same energy.
D. Not enough information to determine

Questions 5–8 refer to the following device: a proton ejector. Follow the directions for each question carefully.

5. A proton ejector device emits 10^3 protons at 10^4 m/s in the same plane as, but at a right angle to, a region of uniform magnetic field strength of 10^{-2} tesla as shown. What is the magnitude of the resulting magnetic force? Assume proton charge = 1.6×10^{-19} C.

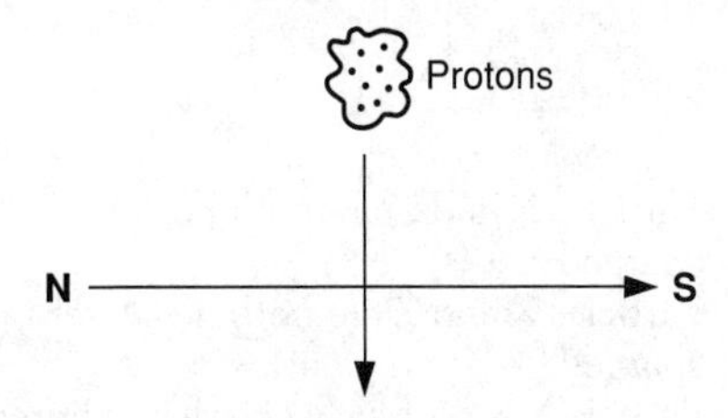

A. 1.6×10^{-10} N
B. 1.6×10^{-14} N
C. 1.6×10^{-17} N
D. 0.8×10^{-17} N

6. What is the direction of the magnetic force produced in question 5?

A. It is oriented directly out of the page.
B. It is oriented directly into the page.
C. It is oriented in the $^{+}$x direction.
D. It is oriented in the −x direction.

7. Based on the information in question 5, which description BEST reflects the change in energy status of the proton burst as it moves through the uniform magnetic field?

   A. Increased
   B. Decreased
   C. No change
   D. Not enough information to determine

8. If the position of the proton ejector device described in question 5 is changed such that the same proton burst now travels from the north to the south pole and in the plane of the existing uniform magnetic field, what is the direction of the magnetic force produced?

   A. It is oriented directly out of the page.
   B. It is oriented directly into the page.
   C. It is oriented in the ^{+}x direction.
   D. None of the above

9. The following diagram represents the paths of three different particles that are traveling in the plane of the paper. A uniform magnetic field passes perpendicularly through the plane and is directed into the plane of the paper. Which statement is most likely to be true?

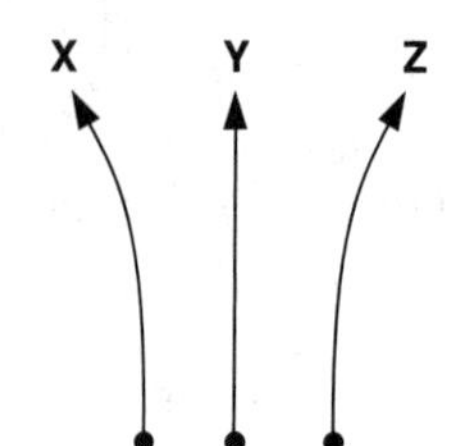

   A. Particles X and Z are both positively charged.
   B. Particles X and Z are both negatively charged.
   C. Particle X is positively charged, whereas particle Z is negatively charged.
   D. Particle X is negatively charged, whereas particle Z is positively charged.

Questions 10 and 11 refer to the following information and diagram.

10. A small metal fragment of mass 30 grams and electrical charge $^{-}4$mC is shot out of an ejector device with an initial horizontal speed of 100 m/s as shown. Assuming that this experiment is conducted on the Earth's surface and within a huge, sealed vacuum chamber, what is expected to happen to the fragment after it leaves the ejector barrel?

    Note that point X is at the same vertical height as the newly released metal fragment, whereas points Y and Z are above and below the horizontal reference line, respectively.

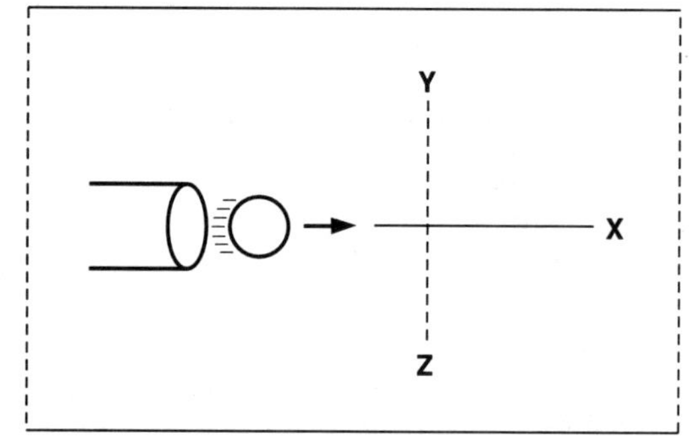

   A. It accelerates in the horizontal direction.
   B. It accelerates in the vertical direction.
   C. It travels purely in the horizontal direction and contacts point X along the right chamber wall.
   D. Both A and C are correct.

11. What are the magnitude and orientation of the magnetic field required to ensure that the metal fragment reaches point X (i.e., there is no change in the metal fragment's vertical component)?

   A. 0.8 tesla, directed into the plane of the paper
   B. 0.8 tesla, directed out of the plane of the paper
   C. 1.3 tesla, directed into the plane of the paper
   D. 1.3 tesla, directed out of the plane of the paper

12. A 1-meter long copper wire is attached between the terminals of a voltage source as shown next. Which diagram correctly depicts the direction of the induced surrounding magnetic field?

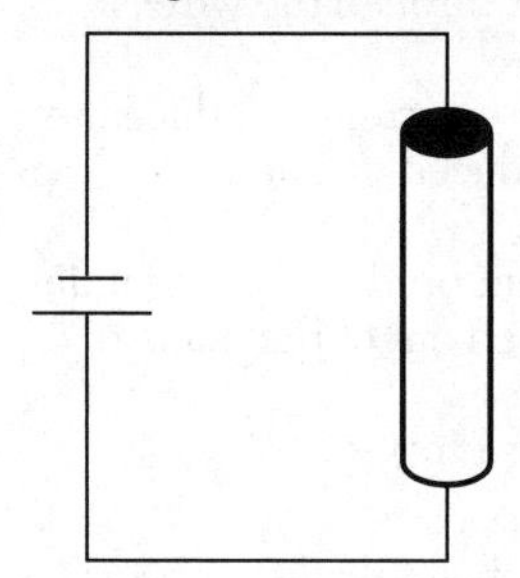

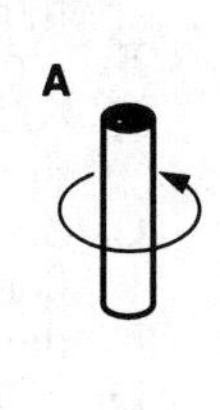

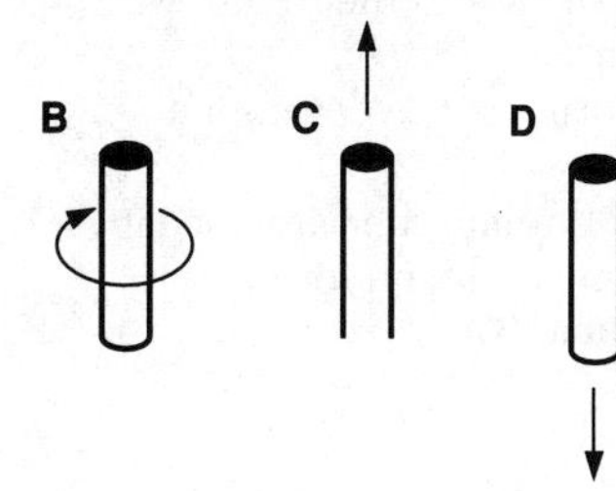

13. Current moving through a wire produces a surrounding magnetic field. If the magnitudes of the magnetic fields produced by wires one, two, and three are equal, which BEST approximates the direction of the net magnetic field at point C?

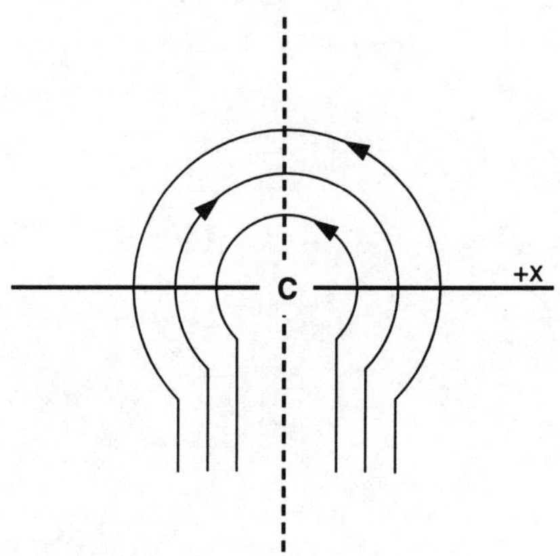

A. Into the plane of the paper
B. Out of the plane of the paper
C. Toward +x
D. Toward −x

Questions 14–19 refer to the following diagram:

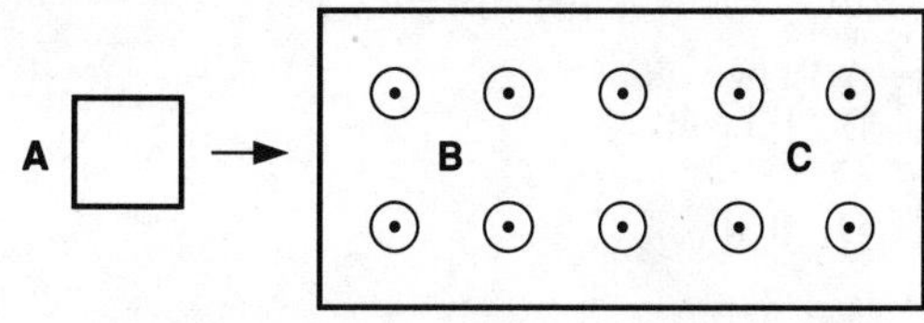

14. A perfectly square metal loop with a side length 5 cm is moved from a distant region "A" toward a vertical plane with a uniform magnetic field of 2.0 T projecting directly from it. The vertical plane can be considered in the plane of this paper, and the magnetic field (represented by "X") projecting directly from the plane of this paper. If the entire square loop is moved quickly from point A to point B, which is just in front of the vertical plane in 0.5 second, what happens?

A. An increase in magnetic flux through the loop occurs.
B. A current is induced in the wire and produces its own magnetic field, which serves to counteract the increase in magnetic flux.
C. The induced magnetic field is directed into the plane of the paper (vertical plane).
D. All of the above are correct.

15. Based on the information in question 14, what direction does the induced current flow?

A. Clockwise around the square wire
B. Counterclockwise around the square wire
C. No induced current
D. Not enough information to determine

16. What is the net flux through the wire once it is entirely in the plane of the uniform magnetic field (at point B)?

A. 0.005 Wb
B. 0.05 Wb
C. 1.0 Wb
D. 5.0 Wb

17. What is the magnitude of the induced EMF when the loop reaches point B?

A. 0.01 volt
B. 1.0 volt
C. 2.0 volts
D. 10.0 volts

18. After it has passed point B, if the wire loop described in question 14 is suddenly compressed so that its area is slightly decreased, which statement is correct?

A. The magnetic flux through the wire is reduced.
B. A counterclockwise current is induced to counteract the reduced magnetic flux through the wire.
C. An induced current produces a magnetic field that is directed out of the plane of the paper.
D. All of the above are correct.

19. If the wire loop described in question 14 is moved from position B to position C within the vertical plane (a distance of 10 cm) in 0.1 sec, what is the magnitude of the induced EMF?

A. 0.01 V
B. 0.05 V
C. 1.0 V
D. 0 V, no induced EMF

SOLUTIONS

DC Circuits and Electromagnetism

1. **D** Power = $V^2/R = (10V)^2/R$

   Before this question can be solved, the resistance must be calculated:

   $1/R = 1/5\ \Omega + 1/3\ \Omega \qquad R = 15/8\ \Omega$

   Plugging the value for R into the expression for power gives:

   $100\ V^2/(15/8\ \Omega) = 160/3\ W$

2. **A** Before the current of the system can be determined, the net resistance of each parallel element of the circuit must be calculated:

   $1/R = 1/3\ \Omega + 1/6\ \Omega \qquad R = 2\ \Omega$

   $1R = 1/4\ \Omega + 1/2\ \Omega \qquad R = 4/3\ \Omega$

   The total resistance of the circuit = R_T = $2\ \Omega + 4/3\ \Omega = 10/3\ \Omega$.

   To find the circuit current, $I = V/R_T$ = $10\ V/(10/3\ \Omega) = 3\ A$.

   Therefore, 3 A flows through the system. To find the current flowing through the 4 Ω resistor, set up a resistance ratio to current ratio chart. Recall that current flows through resistors in parallel and in an inverse ratio to the resistance of each resistor.

Resistance	Resistance ratio	Current ratio	Current flow
4 Ω	2	1	1 A
2 Ω	1	2	2 A

   Note that of the 3 A that enter the parallel component containing the 4 Ω resistor, 2 A go through the 2 Ω resistor and 1 A goes through the 4 Ω resistor.

3. **D** Choices A and C are incorrect. A magnetic force is exerted only if a component of the motion of the charge is perpendicular to the magnetic field. Note that because the magnetic force is always perpendicular to the charge velocity (v), its speed does not change, only its direction.

4. **A** The PA view x-ray has a wavelength of 0.03 nanometer, which converts to 10^{19} Hz via the expression: $f = c/\lambda$, where f = frequency, c = speed of light, and λ = wavelength. The greater the frequency, the greater the EM wave energy.

5. **B** A magnetic force is produced because a charged particle is moving perpendicularly through a magnetic field. The magnetic force (F_m):

   $F_m = qvB\sin 90°$

   $= (10^3\ \text{protons})(1.6 \times 10^{-19}\ \text{C/proton})(10^4\ \text{m/s})(10^{-2}\ \text{T})(1)$

   $= 1.6 \times 10^{-14}\ \text{N}.$

6. **A** The right-hand rule quickly gives the answer to this question. Point the right thumb in the direction of proton travel (downward) and the remaining fingers in the direction of the B-field (to the right); the B-field points to the direction that the right palm faces—out of the plane of the paper, directly at the person following these instructions.

7. **C** A charged particle does not gain or lose energy while moving through a B-field. Some external device or process is required to accelerate a moving particle.

8. **D** A–C are incorrect. Whether the proton burst travels from the north to the south pole or from the south to the north pole, no magnetic force is produced because, by definition, $F_m = qvB\sin 0$, and sin 0° and 180° are both zero. $F_m = 0$ when v is parallel to B and maximum when v is perpendicular to B.

9. **C** Charge "X" is positive, "Y" is neutral, and "Z" is negative. Use the right-hand rule to figure out this question. If the B-field is directed into the page (point fingers into plane of paper) and the direction of movement is up toward the reference points (point right thumb up), the palm gives the direction of the magnetic force on a positively charged particle (toward the right). The back of the right hand gives the direction of the magnetic force on a negatively charged particle (to the left).

10. **B** Gravitational acceleration occurs. Therefore, the acceleration vector points downward in the negative vertical direction.

11. **B** The key concept in this problem is that the metal fragment must continue on the horizontal line drawn without accelerating downward toward position Z (due to gravity). Therefore, a B-field can be placed to counteract the effects of gravity. The forces upward (F_m) equal the forces downward (w). This fact leads to the equation:

$F_m = w$

$qvB = mg$

Therefore, $B = mg/qv = (0.03\text{ kg})(10\text{ m/s}^2)/(4 \times 10^{-3}\text{C})(100\text{ m/s}) = 0.8\text{ T}$.

To figure out the orientation of the magnetic field, use the right-hand rule. Point the back of the hand toward point Y (upward) because it is the direction for the magnetic force to act on the negatively charged fragment. The right thumb points toward point X. The free fingers point directly at the person (out of the plane of the paper).

12. **A** The current (protons, by convention) travels down the (+) terminal and then up the wire. Using the right-hand rule version 2 (see chapter 15), point the right thumb up in the direction of (+) charge flow, and the fingers curl in the direction of the B-field produced. Choice A represents the direction of the B-field surrounding the wire. In addition, recall that the magnitude of the B-field produced is proportional to the magnitude of current. Also remember that the intensity of the surrounding B-fields decreases as one moves farther from the current-carrying wire.

13. **B** Using the right-hand rule version 2, determine the direction of the resulting B-fields. The magnitudes of all three B-fields are equal, and both the inner and outer wire loops have current running in a counterclockwise direction; therefore, the right-hand rule curls two B-fields directed out of the plane of the paper. The net of two B-field units oriented out of the page and one oriented into the page is one B-field unit directed out of the plane ($2\ B_{out} - 1\ B_{in} = 1\ B_{out}$).

14. **D** Answers A, B, and C are all correct statements and step through why an induced current and subsequent induced B-field occur. Remember that any change is counteracted in any way possible. Here, an increase in magnetic flux through the square loop is countered by the induced flow of charge and subsequent induction of an oppositely directed B-field.

15. **A** To counteract the B-field directed out of the paper, the induced current needs to produce a B-field that is directed into the paper. Therefore, the right-hand rule version 2 is used to find out where the induced current flows. One can curl the fingers of the right hand, and the fingertips point into the plane of the paper. Therefore, the induced current travels around the metal loop in a clockwise direction.

16. **A** The flux $= \phi = BA\cos\phi = (2.0\text{ T})(0.0025\text{ m}^2)(1) = 0.005\text{ Wb}$.

17. **A** The magnitude of the average induced $EMF = \Delta\phi/\Delta t = 0.005\text{ Wb}/0.5\text{ sec} = 0.01$ volt. It should be pointed out that the induced EMF is normally assigned a negative value (e.g., $^-0.01$ volt) to emphasize that the induced EMF causes a current that induces a B-field that opposes the changing magnetic flux.

18. **D** Answers A, B, and C are all correct statements and account for what occurs after the change in area of the ring. A decrease in flux occurs because now less area is exposed to the prevailing B-field. To oppose this decrease in magnetic flux, the loop must induce a current that produces a B-field that is directed out of the plane.

19. **D** No change in magnetic flux occurs with the movement of a loop from one region of a uniform B-field to another region within that B-field. Because magnetic status does not change, no induced EMF, current, or B-fields are produced.

HIGH-YIELD REVIEW QUESTIONS

Section II: Physics

Comprehensive Quiz

1. Which statement is true of the adiabatic expansion from b to c?

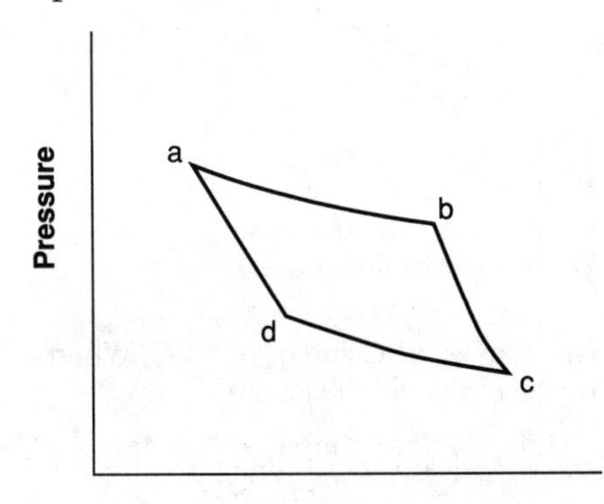

A. The internal energy at b is less than at c.
B. The internal energy at b is greater than at c.
C. The internal energy at b is the same as at c.
D. No comparison can be made based on the data.

2. If the internal energy of a system decreases, which phenomenon can explain this finding?

A. A gas is adiabatically expanded.
B. A gas is compressed by an outside force.
C. A liquid is converted to a gas.
D. Work is done on the system.

3. Three amps of current flow through a 4 Ω resistor in 4 seconds. How much energy is dissipated in this process?

A. 110 J
B. 144 J
C. 48 J
D. 64 J

4. Consider the statement: "Direct contact is not needed for heat transfer." Which processes are consistent with this statement?

I. Conduction
II. Convection
III. Radiation
IV. Sublimation
V. Evaporation

A. I, II, III, IV, and V
B. II, IV, and V
C. III, IV, and V
D. II, III, IV, and V

5. 150 grams of water at 50°C are added to an insulated container holding 1 gram of steam at 220°C. What is the temperature of the system at equilibrium?

Specific heat of water: 1 cal/g°C
Specific heat of steam: 0.5 cal/g°C
Heat of vaporization: 540 cal/g

A. 55°C
B. 69°C
C. 97°C
D. 103°C

6. A magnitude of 4050 calories of heat is added to 150 grams of solid at 10°C. The solid melts at 35°C. The specific heat of the solid is 0.75 cal/g°C and is 0.5 cal/g°C when it is a liquid. Its heat of fusion is 7 cal/g°C. What is its final state and temperature?

A. Liquid between 35 and 45°C
B. Solid between 30 and 40°C
C. Liquid at 25°C
D. Solid at 35°C

7. A system can be taken from state A to B by either of two paths as shown in the following diagram. If the change in internal energy is $^{-}2000$ J, and the work done by the system over path 1 is 600 J, how much heat is lost by the system over path 1?

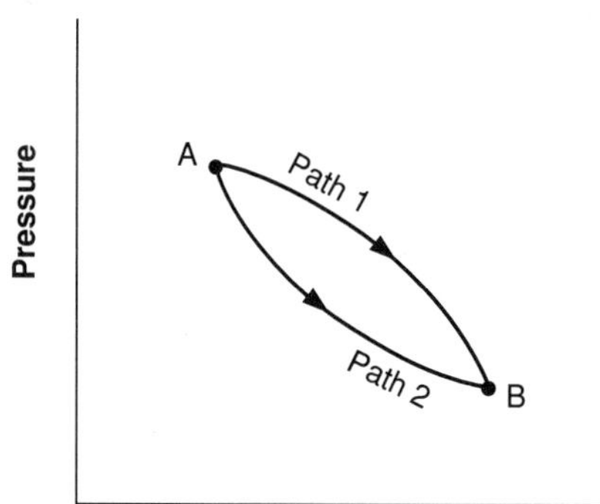

A. 800 J
B. 1400 J
C. 600 J
D. 1000 J

8. Which law of thermodynamics explains the irreversibility of most processes?

A. First law
B. Second law
C. Third law
D. Fourth law

9. At what temperature does a centigrade thermometer show the same value (numeric) as a Fahrenheit thermometer?

A. $^{-}40$
B. $^{-}20$
C. $^{+}20$
D. $^{+}40$

10. During an adiabatic expansion, which statement is true?

A. Internal energy decreases.
B. Work decreases.
C. Heat is removed from the system.
D. Gibbs' free energy decreases.

11. Q_1, Q_2, and Q_3 are three-point charges of equal sign and magnitude. What happens to the force on Q_3 when the sign of Q_2 is reversed?

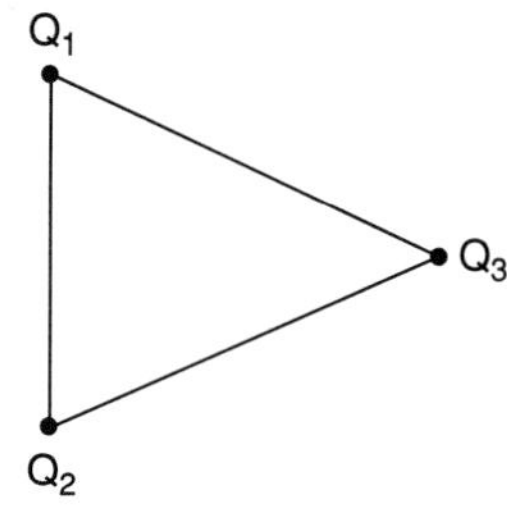

A. It becomes zero.
B. It rotates 90 degrees.
C. It rotates 180 degrees.
D. It reverses direction.

12. Given $q_1 = {}^{+}1$ C and $q_2 = {}^{-}1$ C. Where is the electric field the greatest?

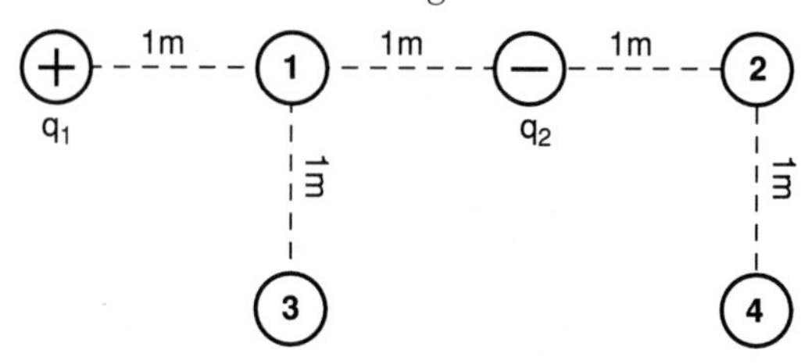

A. 1
B. 2
C. 3
D. 4

13. An unknown electric field exerts a force of 10×10^{-6} N on a $^{-}2 \times 10^{-10}$ C point charge. What is the magnitude of the field at the point charge?

A. 5×10^{2} N/C
B. 5×10^{-4} N/C
C. 5×10^{-2} N/C
D. 5×10^{4} N/C

14. Eight-point charges are arranged in a circle as shown in the diagram. Assume that the radius of the circle is 1 meter, and all charges are $^{+}1\ \mu C$ and equally separated. $k = 9 \times 10^9\ Nm^2/C^2$

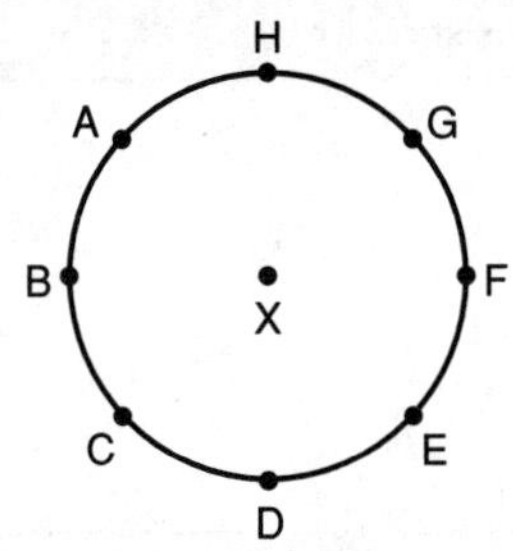

What is the magnitude of the electric field at point X?

A. $180 \times 10^9\ Nm^2/C^2$
B. $90 \times 10^9\ Nm^2/C^2$
C. $45 \times 10^9\ Nm^2/C^2$
D. 0

15. The electric field inside a conductor is:

A. zero.
B. dependent on the charge of the conductor.
C. dependent on the radius of the conductor.
D. variable.

16. The potential difference between two points can be determined by:

A. knowing the relative potential at each point.
B. knowing the electric field magnitude, direction, and the distance between the points.
C. knowing the potential at one point and the difference in position between the points.
D. both A and B.

17. A capacitor carries a charge of X with air between its plates. When a substance of dielectric constant (k = 3) is placed between the plates:

A. the capacitance decreases by a factor of 3.
B. the voltage decreases by a factor of 3.
C. the charge decreases by a factor of 3.
D. the charge decreases by a factor of 6.

18. A parallel plate capacitor with k = 2 is separated by X units with plate area Y units. Assume $E_o = Z$. What is the capacitance?

A. YX/2Z
B. YZ/2X
C. 2YX/Z
D. 2YZ/X

19. Metal rod A has a resistance four times that of rod B. The length of rod A is 4 cm, whereas that of rod B is 2 cm. The radius of rod A is three times that of rod B. What is the ratio of the resistivity of rod A to rod B?

A. 18:1
B. 9:1
C. 24:1
D. 27:1

20. How much current passes through the 2 Ω resistor?

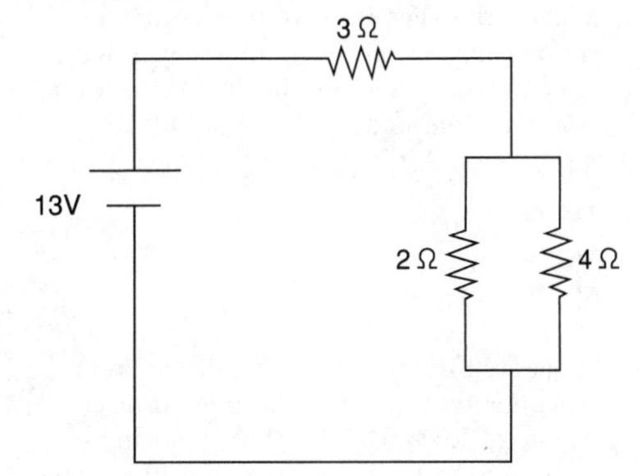

A. 1 amp
B. 3 amp
C. 2 amp
D. 13 amp

21. What is the potential drop from Z to Y?

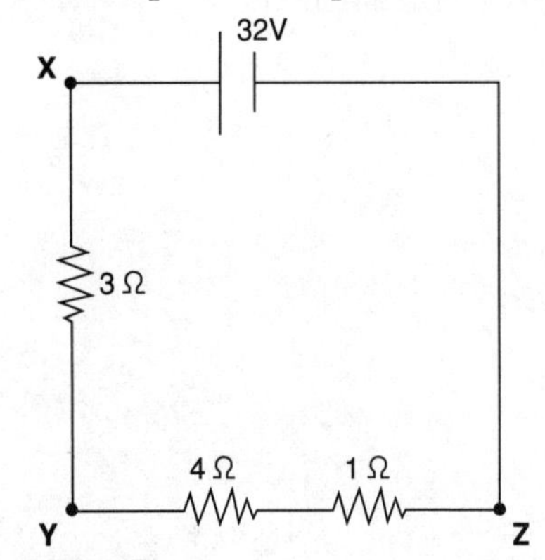

A. 20 volts
B. $^{32}/_{5}$ volts
C. 4 volts
D. 16 volts

22. What is the current passing through the 6 Ω resistor?

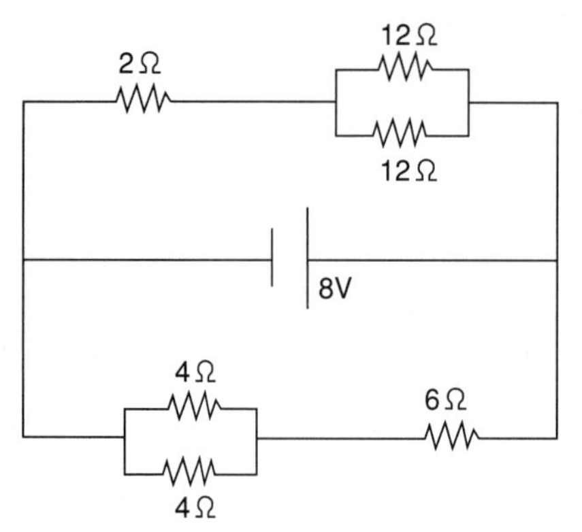

A. 0.5 amp
B. 1 amp
C. 1.7 amp
D. 2.5 amp

23. What is the magnitude of the coefficient of thermal expansion for volume, given that a cube of 8 cm^3 increases its volume by 2 cm^3 when the temperature increases by 10°C?

A. $\frac{1}{20}$
B. $\frac{1}{40}$
C. $\frac{1}{60}$
D. $\frac{1}{80}$

24. What are the magnitude and direction of the magnetic force acting on a single proton charge of 1.6×10^{-19} C that is moving parallel to and in the same plane as a known magnetic field of magnitude 1 tesla at a speed of 4000 m/s?

A. 6.4×10^{-16} N directed out of the plane
B. 4.0×10^{-23} N directed out of the plane
C. 6.4×10^{-16} N directed to the plane
D. None of the above

25. A proton charge (1.6×10^{-19} C) travels at 2.0 $\times$ 10^6 m/s between two magnets as shown in the following diagram. What are the magnitude and direction of the magnetic force if the proton travels 60 degrees from the vertical as shown?

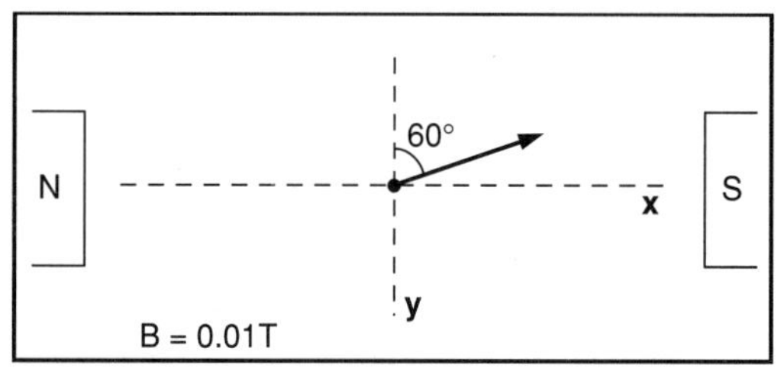

A. 2.8×10^{-15} N directed into the paper
B. 1.6×10^{-15} N directed out of the paper
C. 2.8×10^{-15} N directed out of the paper
D. 1.6×10^{-15} N directed into the paper

26. An electron device releases a sudden burst of 1000 electrons (charge of an electron is 1.6×10^{-19} C) at a speed of 10^8 m/s along the horizontal axis of a coordinate system as shown on the vertical plane in the diagram. A magnetic force of 8×10^{-13} N results and is directed downward along the vertical axis as shown. What are the magnitude and direction of the local magnetic field?

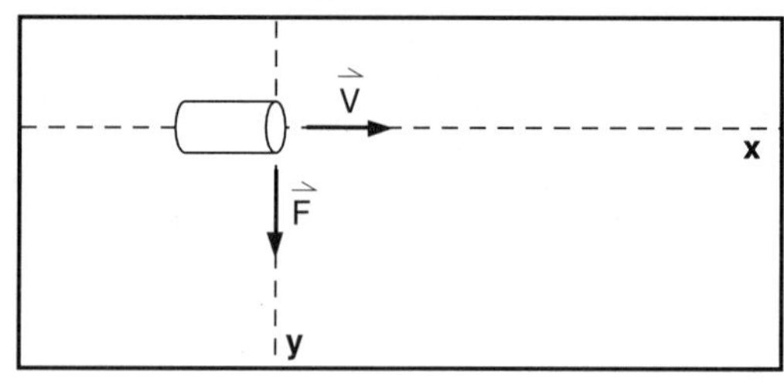

A. 5×10^{-2} tesla directed out of the page
B. 5×10^{-5} tesla directed into the page
C. 5×10^{-2} tesla directed into the page
D. 5×10^{-5} tesla directed out of the page

27. What is the ratio of the magnetic field produced in a single loop wire circuit of radius 50 mm to the magnetic field produced in a 10-loop, 10 cm long solenoid circuit if 0.6 amp of current travels through each circuit?

A. 10:1
B. 1:100
C. 1:10
D. 100:1

28. If a circular loop of wire is moved toward and just in front of a vertical plane with a uniform magnetic field directed into it, what is the direction of the induced current in that loop and the reason for that specific direction?

A. Clockwise on the loop as predicted by Lenz's law
B. Clockwise on the loop as predicted by Faraday's law
C. Counterclockwise on the loop; Lenz's law
D. Counterclockwise on the loop; Faraday's law

29. A perfectly circular wire with two complete loops and a diameter of 20 cm is mounted vertically in a holding brace. Next, a device capable of producing numerous magnetic field strengths is set up adjacent to the wire loop so that magnetic field lines generated pass directly through the loop's open central portion (lines are perpendicular to the plane of the loop). What is the change in magnetic flux when the magnitude of the magnetic field increases from 1.4 T to 3.4 T?

A. 0.06 Wb
B. 0.25 Wb
C. 10 Wb
D. 40 Wb

SOLUTIONS

Section II: Comprehensive Quiz

1. **B** Because the process of (b) to (c) is an adiabatic expansion, $\Delta Q = 0$. Since $\Delta U = \Delta Q - w$, $\Delta U = 0 - w = {}^{-}w$. If P decreases while volume increases, work must have been done by the system. Therefore, the system's internal energy must be greater at point (b) than at point (c).

2. **A** The internal energy of a system decreases when a gas is adiabatically expanded as work is being done by the system. Choices B–D are incorrect because they involve increasing ΔU.

3. **B** $P = I^2R$. This equation gives power in units of watts (joules/sec). The answer choices give units of energy (joules). Multiply by time required; the answer is in joules. Energy dissipated = $(3\text{ A})^2(4\text{ ohms})(4\text{ sec}) = 144$ joules.

4. **D** Although convection, radiation, sublimation, and evaporation do not require direct contact for heat transfer, conduction does, by definition.

5. **A** Assume that heat loss by steam = heat gain by water. Assume that the system comes to rest as a liquid at a temperature greater than 50°C and less than 100°C. That equilibrium temperature can be called T_f. Therefore: $(1\text{ g})(0.5\text{ cal/g°C})(220°C - 100°C) + (1\text{ g})(540\text{ cal/g}) + (1\text{ g})(1\text{ cal/g°C})(100°C - T_f) = (150\text{ g})(1\text{ cal/g°C})(T_f - 50°C)$. Steam cooling to 100°C + steam condensing + water cooling = water originally at 50°C warming.

   Solving the equation for T_f gives a final system temperature of 55°C.

6. **A** This difficult question requires an "energy budget" technique. First, 4050 cal of heat is added to the solid object. How much energy does it take to melt this object? Because $Q = mc\Delta T$, $Q = (150\text{g})(0.75\text{ cal/g°C})(35°C - 10°C) = 2813$ cal required just to get the solid to its melting point.

   To melt the solid requires a phase change: $(150\text{ g})(7\text{ cal/g}) = 1050$ cal.

   2813 cal + 1050 cal = 3863 cal used to warm and melt the solid. Because 4050 cal are available and only 3863 cal have been used, 187 cal (4050 − 3863) are available for additional warming.

   Therefore, $(150\text{ g})(0.5\text{ cal/g°C})(T_f - 35°C) = 187$ cal

   $T_f = 37.5°C$

7. **B** $\Delta U = \Delta Q - w$.

   ${}^{-}2000 = \Delta Q - (600)$
   Solving for ΔQ gives $\Delta Q = {}^{-}1400$ J. This solution implies that 1400 joules were lost by the system over path 1.

8. **B** The second law of thermodynamics says that ΔS (entropy) tends to increase naturally. Recall that entropy means disorder. Therefore, most spontaneous and entropic thermodynamic processes are irreversible.

9. **A** Recall that $T_f = \frac{9}{5}T_c + 32$. Plug in values from each choice into this equation to see which is best. With choice A, $T_f = \frac{9}{5}({}^{-}40°C) + 32 = {}^{-}40°F$.

10. **A** $\Delta Q = 0$ during adiabatic expansion; therefore, $\Delta U = \Delta Q - w = 0 - w$, so ΔU decreases. Work is being done by the system.

11. **B** First, draw a diagram that shows the vector that results from the given charges:

1

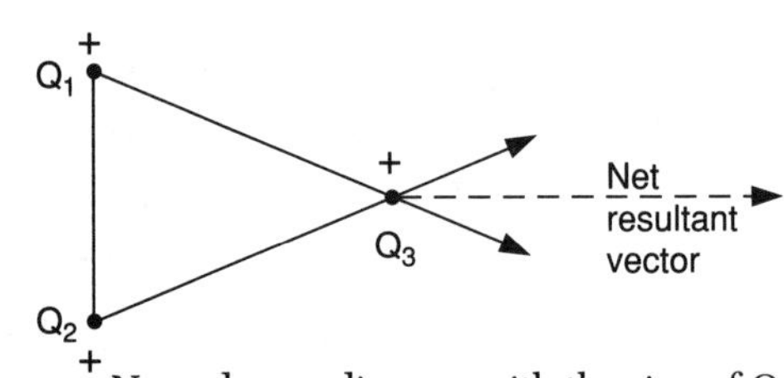

Now, draw a diagram with the sign of Q_2 reversed. Note that the resultant vector shifts downward 90° compared to the original resultant vector.

2

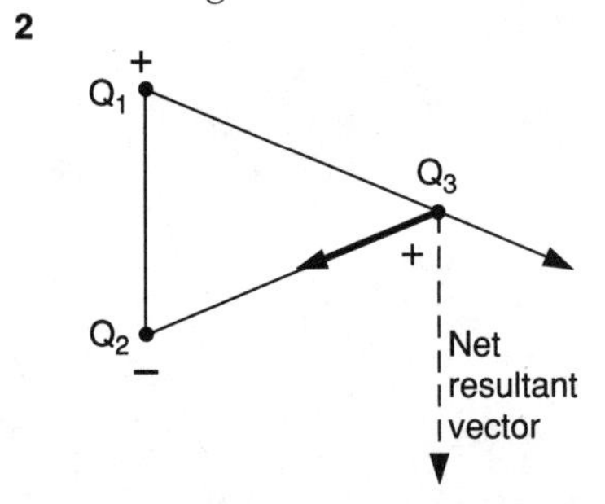

12. **A** An important concept is shown in this question: as a small positive point charge moves away from a charge (q), the E-field that it experiences decreases because $E = kQ/r^2$. Note that the r term is in the denominator. Quantitatively:

At point 1: $E = kq_1/1^2 + kq_2/1^2 = 2\ kq$

At point 2: $E = kq_1/3^2 + kq_2/1^2 = (^{10}\!/_9)kq$

Note that the closer the point of interest is to the (q) charges, the greater the E-field. Therefore, points 3 and 4 have smaller E-fields than point 1.

13. **D** $E = F/q = (10 \times 10^{-6}\ N)/(2 \times 10^{-10}\ C) = 5 \times 10^4\ N/C$.

14. **D** Note that this situation in which the E-field vectors cancel is symmetrical. This can be shown by putting a positive test charge at point X. The net magnitude of the E-field is zero at point X.

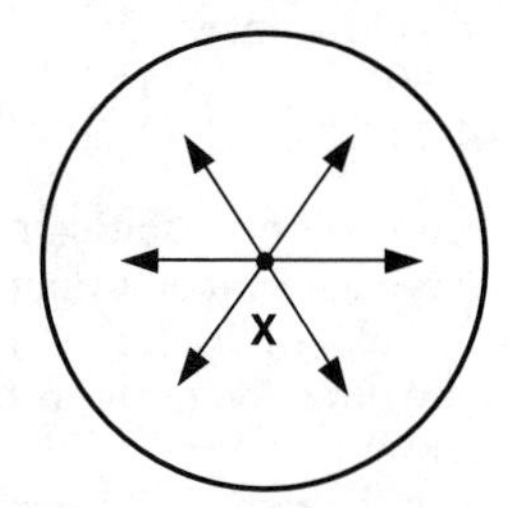

15. **A** Recall that the E-field inside a conductor is zero.

16. **D** Both A and B are true statements, whereas choice C is incorrect. Recall that $\Delta V = V_B - V_A = \Delta PE/q$. In addition, remember that $\Delta V = Es$, where E is a uniform E-field and (s) is the distance between two points, A and B.

17. **B** If a dielectric is placed within a capacitor carrying a charge (q constant), $C_{new} = kC_{initial}$. Therefore, the capacitance goes up by three times. Because $Q = CV$ and Q is constant, the voltage must decrease by a factor of 3.

18. **D** To solve this question, substitute the given values into the expression for capacitance: $C = kE(A/d)$, where k is a dielectric constant, E is a constant, A is area of a capacitor plate, and d is the distance between the plates.

Plugging in values from the question shows that $C = 2ZY/X$.

19. **A** It is known that $R = (1/A)\rho$; therefore, $\rho = RA/1$.

$\rho_A = (4)(3^2)/(2) = 18$
$\rho_B = (1)(1^2)/(1) = 1$

Therefore, $\rho_A/\rho_B = 18/1$.

20. **C** Solve this problem by first calculating the total circuit resistance.

For the parallel resistors: $1/R_T = 1/2 + 1/4$
$R_T = 4/3\ \Omega$

For the total circuit resistance, add the series elements. Therefore, $R_{circuit} = 4/3\ \Omega + 3\ \Omega = 13/3\ \Omega$

Now, find the total circuit current: $I = V/R = 13\ V/(13/3\ \Omega) = 3\ A$.

To find how much current flows through the 2 ohm resistor, set up a resistance/current table. Recall that current flows through resistors in parallel, in an inverse ratio to the resistances.

Resistance	Resistance ratio	Current ratio	Current
$2\ \Omega$	1	2	$(2/3)(3\ A) = 2\ A$
$4\ \Omega$	2	1	$(1/3)(3\ A) = 1\ A$

Therefore, 3 A flows through the resistors in a 2:1 ratio inversely to their resistances; that is, 2 A flows through the $2\ \Omega$ resistor.

21. **A** Start this problem by finding out how much current flows through the circuit. $V = IR$; therefore, $I = V/R$. Because the resistors are in series, the total circuit resistance is simply the sum of the individual resistances: $R = (3\ \Omega + 4\ \Omega + 1\ \Omega) = 8\ \Omega$ $\quad I = 32\ V/8\ \Omega = 4\ A$. The voltage drop from point Z to Y is the resistance between these two points ($4\ \Omega + 1\ \Omega = 5\ \Omega$) multiplied by the circuit current. $V_{drop} = (4\ A)(5\ \Omega) = 20$ volts.

22. **B** The best way to solve this problem is by reducing the complex circuit given into a simpler one. First, reduce the parallel resistances so that the circuit contains only series elements:

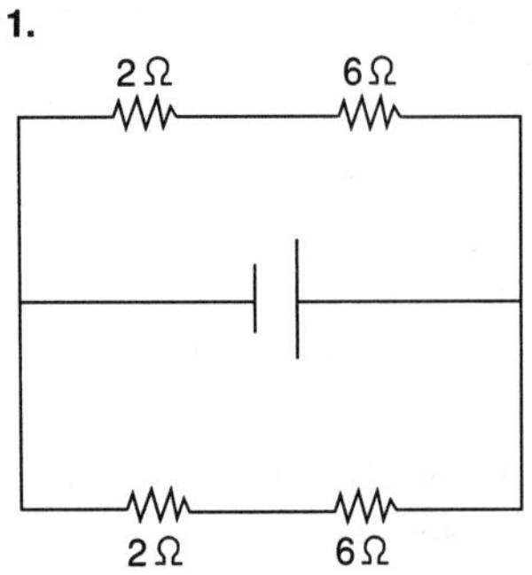

Second, reduce the series resistors by adding them together.

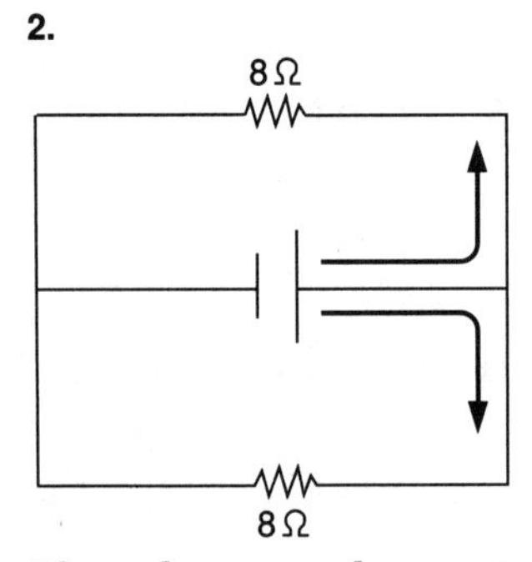

This reduction produces a circuit with two parallel components. Note that as current leaves the battery, it can choose between two paths. Another way to write this circuit is shown next.

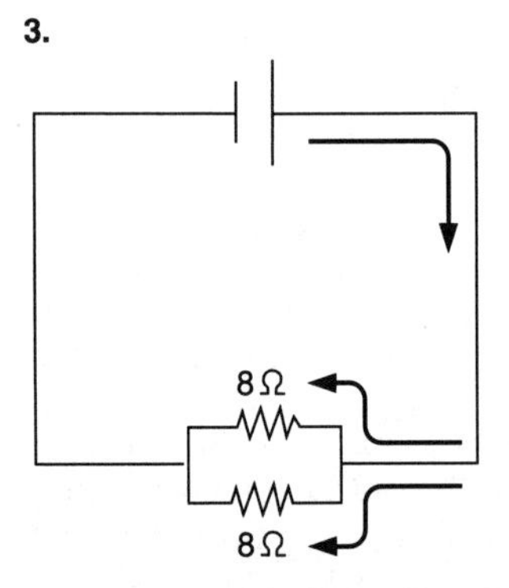

Because V = IR, I = V/R = 8 V/4 Ω = 2 A, the 2 A flow through the whole circuit. Both resistances are the same (8 Ω); therefore, only half of the total current flows through the lower tier of this circuit. In other words, half the current flows through the 6 Ω resistor shown in the original diagram of the circuit (see question 22).

23. **B** Recall that the change in volume associated with thermal change is $\Delta V = \beta V_o(\Delta T)$. If this equation is rearranged, $\beta = (\Delta V/V_o)\Delta T$. Therefore, $\beta = (2)/(8)(10) = 1/40$.

24. **D** By definition, a charge that is moving parallel to the magnetic field experiences no magnetic force. This definition holds true for charges traveling in the same direction as the B-field (because the sin 0° equals zero) and for charges traveling in the opposite direction of the B-field (because the sin 180° equal zero). Mathematically, $F_{mag} = (1.6 \times 10^{-19}$ C)(4000 m/s)(1 T)(sin 0°) = 0.

25. **D** This question is a fairly straightforward magnetic force problem. However, note that the moving proton makes a 30° angle with the horizontal axis. The 60° angle with the vertical axis is not used for further equation solving. Therefore, the magnitude of the magnetic force is: $F_{mag} = qvB\sin 30° = (1.6 \times 10^{-19}$ C)$(2 \times 10^6$ m/s)(0.01 T)(0.5) $= 1.6 \times 10^{-15}$ N.

The direction of the resulting magnetic force can be determined by applying the right-hand rule. In brief, the right fingers point in the direction of the B-field present (from the north pole to the south pole, i.e., to the right in this problem), and the thumb points in the direction of the particle's velocity vector (elevated 30° above the horizontal axis in this situation). If the moving particle in question has a positive charge, the resulting magnetic force points in the direction of the open right palm (the open right palm faces into the paper given the direction of v and the B-field in the problem). Remember that for negatively charged particles, the resulting magnetic force is exactly opposite the direction of the open palm; that is, the force points toward the back of the right hand.

26. **B** The magnitude of the B-field can be quickly determined using the following equation: $B = F/qv\sin\theta$

Therefore, $(8 \times 10^{-13}$ N)/(1000)$(1.6 \times 10^{-19}$ C)(108 m/s)(1) $= 5 \times 10^{-5}$ tesla.

The direction of the B-field can be found if the right thumb is pointed to the right (toward +x) and the back of the right hand is aimed downward (toward −y). This position shows that the B-field points directly into the plane of the paper. Note that this problem involves moving electrons; therefore, aim the back of the right hand downward in the $^{-}$y direction. Use the palm to determine an unknown F_{mag} or to point toward an already known F_{mag} only when dealing with a positively charged particle.

27. **C** Knowing the formulas for the magnitude of the B-field produced by each type of circuit allows you to set up a ratio that quickly leads to the answer. Be careful when it comes time to do the calculation, because units are not equivalent in this problem. In ratio-type problems, it is critical to have all like-measure numbers in the same unit (in this situation, all lengths in either cm or meters).

$(B_{loop})/(B_{solenoid}) = (\mu I/2a)/(\mu NI/L) = L/2aN = 1/10$.

28. **C** As Lenz's law suggests, the effect of moving the loop toward a B-field must be countered by an induced current in the loop itself, which subsequently produces an oppositely directed B-field. This "induced" B-field counters the effect of the initial action. If the vertical plane has a B-field directed right into it, the induced current is in a direction that produces a B-field that directly opposes the original inward B-field (i.e., the induced current creates a B-field directed right out of the plane). A current traveling counterclockwise on a circular loop creates a B-field oriented directly out of the plane under study. This field serves to counterbalance the increasing magnetic flux that the loop experiences.

29. **A** Recall the equation for magnetic flux:

$\Phi = BA\cos\phi$, where B is the magnitude of the B-field, A = the area of the loop, and ϕ is the angle that the field makes with the normal to the loop.

The change in flux can be determined by:

$\Delta\Phi = (\Delta B)(\text{area of loop}) = (2T)(\pi r^2)$

In this question, be cautious. First, note that the diameter of the circular wire is given, not the radius. Divide the diameter by two to get the radius. Next, recall that magnetic equations require lengths to be in meters. Therefore, convert the radius (10 cm) into meters (0.10 m). Now, complete the problem.

$\Delta\Phi = (2T)(3.14)(0.1)^2 = 0.06$ Wb.

HIGH-YIELD REVIEW QUESTIONS

Section III: Physics

Waves, Light, Sound, and Nuclear Structure

1. When light enters a medium of refractive index greater than air, which element decreases?

A. f
B. λ
C. f, λ
D. c, f

2. Light travels 1.24×10^8 m/s in a diamond. What is the index of refraction of a diamond?

A. 0.33
B. 1.67
C. 2.33
D. 2.42

3. An underwater light source causes light to hit the water–air interface. The light ray makes an angle θ_3 in air with the normal, and θ_4 in air with the interface plane. How does θ_1 compare to θ_3?

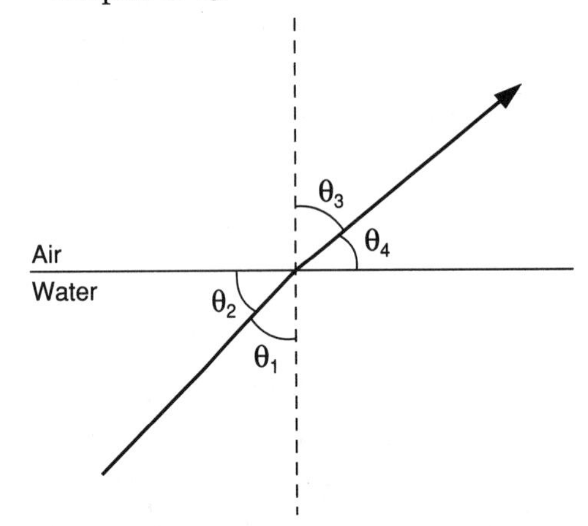

A. $\theta_1 = \theta_3$
B. $\theta_1 > \theta_3$
C. $\theta_1 < \theta_3$
D. No comparison can be made.

4. Light moves from water to air as shown in the following diagram. What (approximate) angle of incidence (θ_1) must light approach to refract light along the interface?

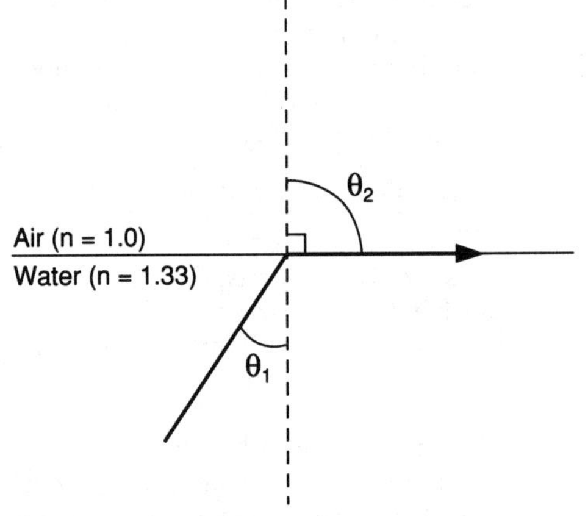

A. 30 degrees
B. 60 degrees
C. 15 degrees
D. 45 degrees

5. In question 4, if θ_1 were greater than the answer to that question, where would the light reflect?

A. Reflect into the water
B. Reflect into the air
C. Reflect along the interface
D. Diffract in both air and water

6. An object is placed between the center of curvature and the focal point of a concave mirror. The image is:

A. real, inverted, enlarged.
B. virtual, upright, enlarged.
C. real, upright, reduced.
D. virtual, inverted, reduced.

7. A man, 6 feet tall, stands 6 meters in front of a convex mirror. The mirror has a center of curvature of 10 meters. Approximately how much is the man's image magnified, and is it real or virtual?

A. 2, real
B. ½, real
C. 2, virtual
D. ½, virtual

8. For convex mirrors, images are always:

A. diminished, upright, virtual.
B. enlarged, upright, virtual.
C. diminished, inverted, real.
D. enlarged, inverted, real.

9. As an object is moved from a great distance toward the focal point of a concave mirror, the image moves from:

A. a great distance toward the focal point and is always real.
B. the focal point toward a great distance from the mirror and is always real.
C. the focal point to a great distance and is always virtual.
D. the focal point to a position immediately adjacent to the mirror and is always real.

10. A lens of focal length $^-50$ cm has what power in diopters?

A. $^-50$
B. $^-2$
C. $^+5$
D. $^+20$

11. An object is placed 10 cm from a diverging lens of focal length $^-20$ cm. The image formed is:

A. real, inverted, reduced.
B. real, inverted, enlarged.
C. virtual, upright, enlarged.
D. virtual, inverted, reduced.

---

12. A convex mirror with focal length F and an object are positioned as shown in the following diagram. The image forms at which letter (A–D)?

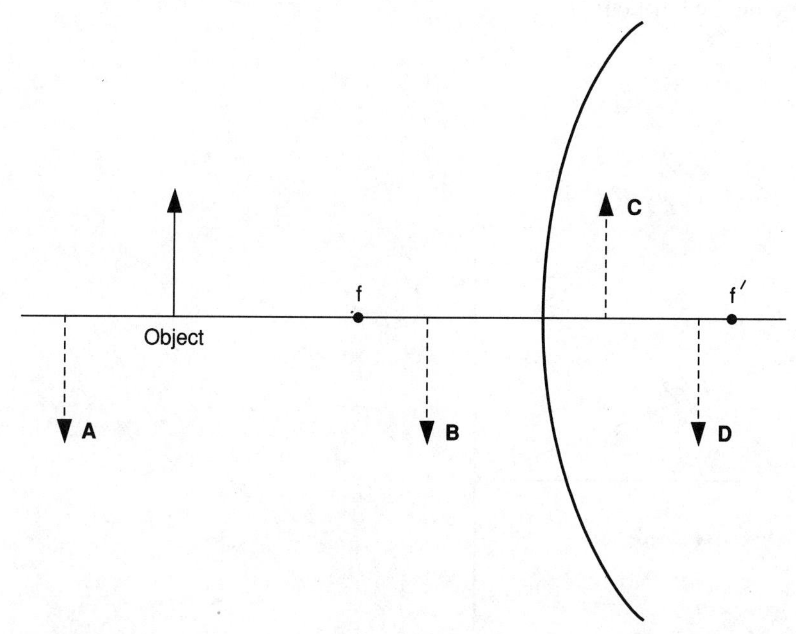

A. A
B. B
C. C
D. D

13. A lens of f = 20 cm and a lens of f = 33 cm are placed 150 cm apart. An object forms an image that forms a second image 1 meter from the second lens. What is the distance from the object to the first lens?

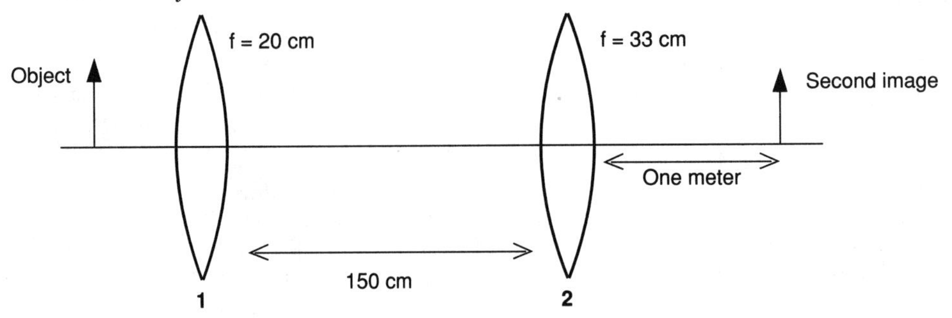

A. 33 cm
B. 25 cm
C. 30 cm
D. 40 cm

14. A lens with a 6-meter focal length and a lens with a 3-meter focal length are placed in contact. Assume that the lenses are thin. What is the focal length of the combination?

A. 9 meters
B. ⅙ meter
C. 12 meters
D. 2 meters

15. If the middle slit is closed, the intensity of light at point A on the screen:

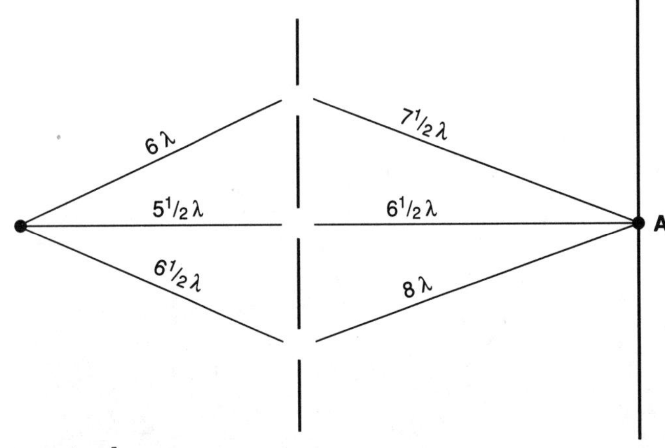

A. decreases.
B. increases.
C. remains the same.
D. cannot be determined.

16. Light passes through three successive media. Light refracts at each interface. In which medium does light travel fastest?

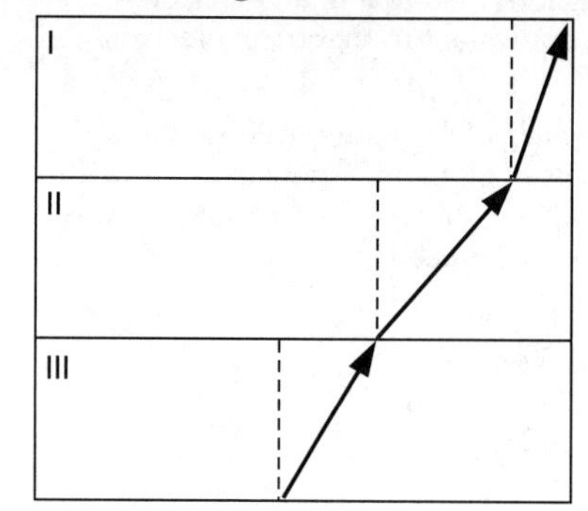

A. I
B. II
C. III
D. Light travels at the same speed in all three media.

17. Two light sources of equal power, P and Q, are 180 degrees out of phase at point R. A third light source, S, is introduced between P and Q. What is their intensity?

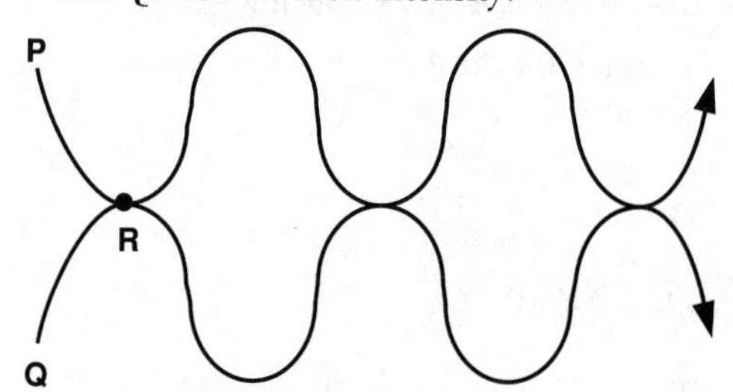

A. The intensity is the average of all three power values.
B. The intensity is the sum of all three power values.
C. The intensity is the sum of P and S power values.
D. The intensity is the same as if P and Q are not present.

18. A sound wave has v = 10,000 ft/sec and f = 1000 Hz. The sound wave is emitted by a source at rest. When the source is moving at constant velocity of 2000 ft/sec, what is the ratio of the f heard by a stationary observer behind the moving source to the frequency heard by an observer in front of the moving source?

A. 1:1
B. 2:3
C. 3:2
D. 1:2

19. What is the wavelength of a sound wave in the air at 0 degrees centigrade if the frequency of the sound is 440 Hz? (velocity of sound in air = 330 m/s)

A. 1.0 m
B. 0.5 m
C. 0.25 m
D. 0.75 m

20. A traveling wave passes a point of observation. At this point the time interval between successive crests is 0.2 sec. Which statement is true?

A. The wavelength is 5 m.
B. The frequency is 5 Hz.
C. The velocity of propagation is 5 m/s.
D. The wavelength is 0.2 m.

21. An instrument has two strings. Both have the same length and are uniform. One of the strings is under twice as much tension and has twice the mass of the other. The strings are plucked. Which statement is true?

A. Both strings have the same wave velocity.
B. The string with greater mass and tension has greater velocity.
C. The string with lesser mass and tension has lesser velocity.
D. None of the statements are true.

22. A periodic simple harmonic oscillator is given by: $y = 3\sin(\pi t)$ meters, where t is in seconds. The period of the system is:

A. 2 s.
B. 2 Hz.
C. 3 m.
D. 0.5 s.

23. Pendulum A has twice the amplitude of swing and twice the mass of pendulum B. Both pendulums have the same length. How does the period of A compare to that of B?

A. $A > B$.
B. $A < B$.
C. $A = B$.
D. The two cannot be compared.

has two hollow tubes closed at one ith lengths unknown. He also has two forks, both of which emit a 10-cm ength. He strikes the forks and places one over each tube. What length should the tubes be for resonance to occur?

A. 2.5 cm and 7.5 cm
B. 5.0 cm and 10.0 cm
C. 7.5 cm and 10.0 cm
D. None of the above

25. A spring is sent into oscillation. If the displacement of the spring from its equilibrium point is measured over time, what type of relationship would exist between displacement and time?

A. Linear
B. Parabolic
C. Sinusoidal
D. Circular

26. A guitar has two strings that are set into vibration. The frequency of string X is 50 cycles/s. The frequency of string Y is 30 cycles/s. What is the beat frequency of these two strings?

A. 80 cycles/s
B. 20 cycles/s
C. 40 cycles/s
D. 100 cycles/s

27. A whistle blows 10 m from a power sensing device. The intensity of the sound is proportional to:

A. $0.1\ m^4$.
B. $0.01\ m^{-2}$.
C. $10\ m^1$.
D. $100\ m^2$.

28. Sound can travel through many different types of media. Sound travels slowest in:

A. air.
B. water.
C. iron.
D. granite.

29. The element ${}^{32}_{16}X$ is formed as a result of alpha and beta decay. Assuming that element Y is the parent compound, what is element Y's atomic number and mass?

A. 35, 16
B. 36, 17
C. 36, 15
D. 35, 15

30. The mass of a gamma particle is:

A. less than that of an electron.
B. more than that of an electron.
C. equivalent to that of an electron.

31. The half-life of element X is 6 days. If 3 grams of element X remain after 30 days of decaying, how much of element X was present originally?

A. 48 grams
B. 96 grams
C. 24 grams
D. 192 grams

32. Compared to the combined mass of its protons and neutrons, an individual nucleus of an atom has a mass that is:

A. smaller.
B. larger.
C. equivalent.
D. not possible to determine.

33. If the binding energy of a nucleus is 186.2 MeV, what is its mass deficit?

1 amu = 931 MeV
1 amu = 1.66×10^{-24} g

A. 2.5×10^{-24} g
B. 3.0×10^{-25} g
C. 3.2×10^{-26} g
D. 3.3×10^{-25} g

34. The most stable elements have the ______ binding energies. This energy arises from the ______ of the nuclear particles. The two words that BEST complete the two statements are:

A. highest, mass.
B. lowest, charge.
C. highest, size.
D. lowest, size.

35. Which statement about the photoelectric effect is false?

A. High-frequency light striking metals can cause electrons to be emitted.
B. The work function is the minimum energy necessary to pry an electron loose.
C. The KE of emitted electrons equals the energy of the incident light waves.
D. The frequency of the incoming light times Planck's constant equals the energy of light protons striking the metal.

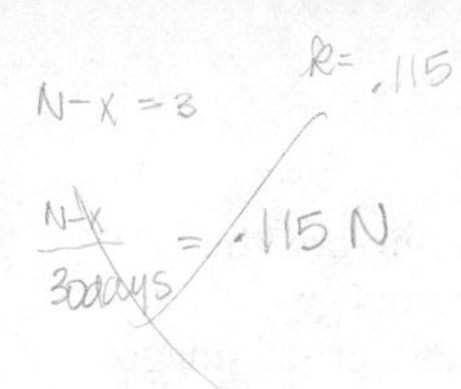

SOLUTIONS

Waves, Light, Sound, and Nuclear Structure

1. **B** Frequency is constant as light passes through media of different n values. However, $\lambda_{medium} = (\lambda_{vacuum})/n$, λ decreases in this problem as (n) increases. The speed of light (c) is a constant, and in a vacuum, $c = 3.0 \times 10^8$ m/s.

2. **D** $n = c/v = (3.0 \times 10^8 \text{ m/s})/(1.24 \times 10^8 \text{ m/s}) = 2.42$.

3. **C** In this problem, light bends away from the normal as it enters a lower n value (air) from a higher n value (water). Light bends toward the normal if it moves from air to water. The following diagram shows light bending away from the normal as it enters a medium of lower n value.

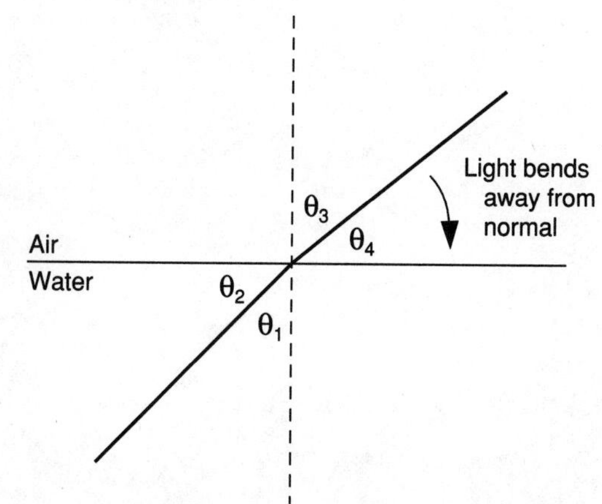

4. **D** This question asks for the critical angle, which is θ_2 or 90 degrees. Recall Snell's law, which states: $n_1 \sin \theta_2 = n_2 \sin \theta_2$.

Therefore, $(1.33)(\sin \theta_1) = (1.0)(\sin 90°)$

$$\sin \theta_1 = (1.0)/(1.33) = 0.75$$

$$\sin^{-1}(0.75) \approx 45°$$

5. **A** If the critical angle for θ_1 (which is approximately 45 degrees) is exceeded, the result is total internal reflection. The light rays are reflected back into the water. The following diagram shows this phenomenon:

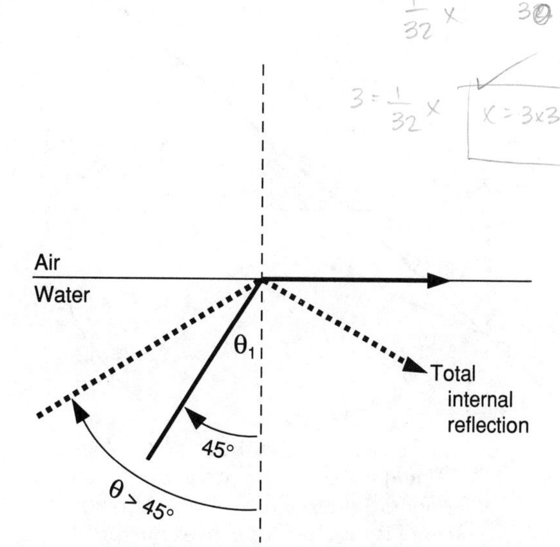

6. **A** This question can be solved either by math equations or by ray diagrams. To solve mathematically, make up values for f and p that follow the description given in the question. Suppose that f = 1 and $p = \frac{3}{2}$.

Therefore, $1/p + 1/q = 1/f$.

$$1/(\tfrac{3}{2}) + 1/q = 1/1$$

$$\tfrac{2}{3} + 1/q = 1$$

$$1/q = \tfrac{1}{3}$$

$$q = 3.$$

Because q is positive, the image is real. Magnification (m) $= -q/p = -3/(\tfrac{3}{2}) = -2$. Because m is negative, the object is inverted. Because the absolute value of m > 1, the image is enlarged.

By the following ray diagram, the object is real (same side of the mirror as the object), enlarged, and inverted:

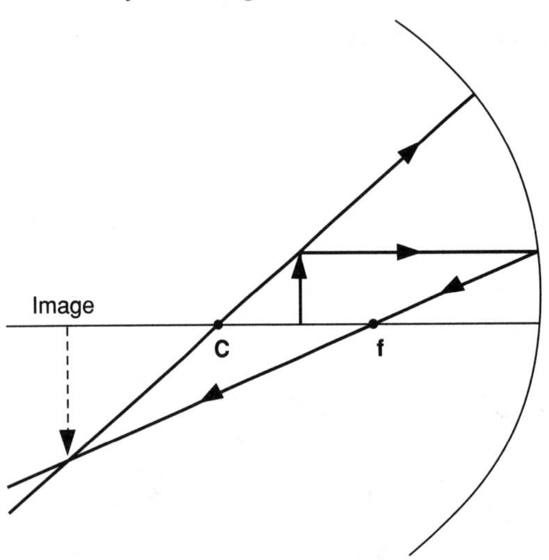

7. **D** For convex mirrors draw the focal length (f) and the center of curvature (C) on opposite sides of the mirror. Also note that f is negative for convex mirrors. By the thin lens equation:

$$1/p + 1/q = 1/f$$

$$\frac{1}{6} + 1/q = \frac{1}{-5}$$

$$1/q = \frac{-11}{30}$$

$$q = \frac{-30}{11}$$

$$m = -q/p = \left(\frac{30}{11}\right)/6 \approx \frac{1}{2}$$

Because q is negative, the image is virtual. Because m is positive, the image is upright. Because $|m| < 1$, the image is diminished.

By the following ray diagram , the image can also be found:

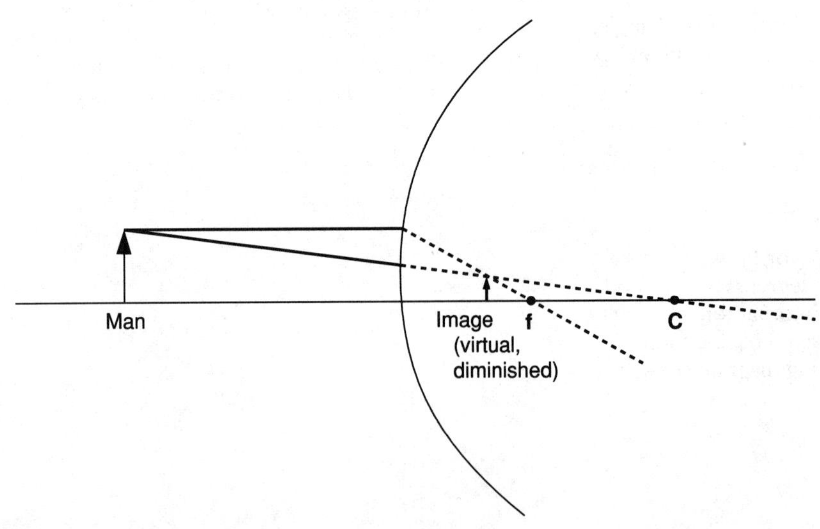

8. **A** Recall that for convex mirrors, the DUV rule holds: Convex mirrors give diminished, upright, and virtual images.

9. **B** This problem can be solved by the ray diagram technique or the equation technique. To solve by the equation technique, make up values for p and f and solve for q. To solve by the ray diagram technique, use the following diagrams to predict image movement.

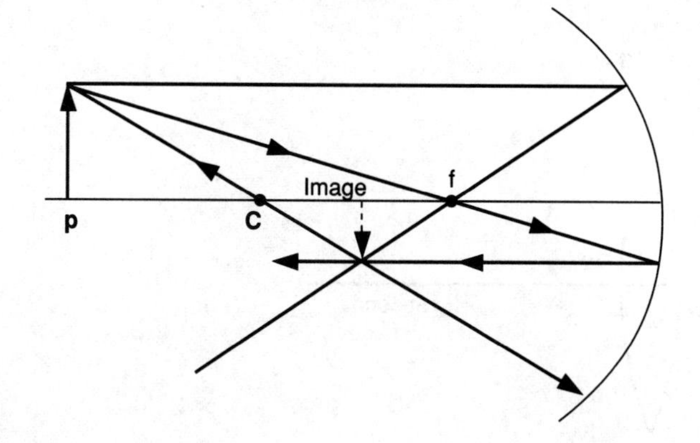

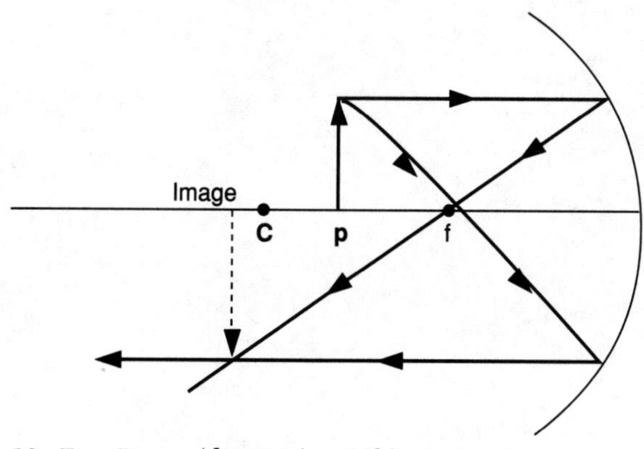

10. **B** Power (diopters) = 1/f (in meters) = 1/(⁻0.5 m) = ⁻2.

11. **C** Convex mirrors and diverging lenses produce diminished, upright, and virtual images, known as the DUV rule. To calculate the answer, use the thin lens equation: $1/p + 1/q = 1/f$

$$\tfrac{1}{10} + 1/q = -\tfrac{1}{20}$$
$$1/q = -\tfrac{3}{20}$$
$$q = -\tfrac{20}{3}$$
$$m = -q/p = (\tfrac{20}{3})/10 = \tfrac{2}{3}$$

Because q is negative, the image is virtual. Because $|m| < 1$, the image is diminished. Because m is positive, the image is upright.

12. **C** Convex mirrors give diminished, upright, and virtual images. Only choice C is consistent with this fact.

13. **B** This challenging problem is known as a combination of lenses problem. Use the diagram of both lenses. Understand that the image formed by the first lens is the object for the second lens. Because the problem gives only the second image distance, one has to work backward to find the location of the first image. Once the first image distance is known, the original object distance can be found. Look at the following diagram:

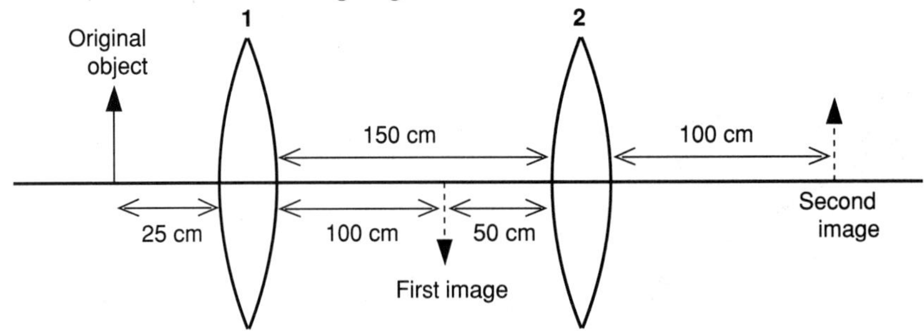

(Note: not drawn to scale)

Start with lens 2. Where is the image of the original object that made the object for the second image?

$$1/p + 1/q = 1/f$$

$$1/p + \tfrac{1}{100}\text{ cm} = \tfrac{1}{33}\text{ cm}$$

$$p = 50\text{ cm}$$

Therefore, the image from the original object is (150 − 50) or 100 cm from lens 1, or 50 cm from lens 2. Now, find p using q = 100 cm and f = 20 cm for lens 1.

$$1/p + \tfrac{1}{100} = \tfrac{1}{20}$$

$$p = 25\text{ cm}$$

14. **D** For two thin lenses in contact, the effective focal length is equivalent to:

$$1/F = 1/f_1 + 1/f_2$$

$$1/F = \tfrac{1}{6} + \tfrac{1}{3}$$

$$1/F = \tfrac{1}{2}$$

$$F = 2\text{ m}$$

The power (diopters) of two lenses is the sum of the power of each lens:

$$D_{total} = D_1 + D_2$$

15. **B** Waves in phase are additive, whereas waves out of phase are destructive and interfere.

The top slit has a total wavelength of: $6\ \lambda + 7\tfrac{1}{2}\ \lambda = 13\tfrac{1}{2}\ \lambda$

The middle slit has a total wavelength of: $5\tfrac{1}{2}\ \lambda + 6\tfrac{1}{2}\ \lambda = 12\ \lambda$

The bottom slit has a total wavelength of: $6\tfrac{1}{2}\ \lambda + 8\ \lambda = 14\tfrac{1}{2}\ \lambda$

Notice that the top and bottom slits have wavelength sums $1\ \lambda$ apart. Therefore, they are in phase and are additive. When the destructive influence of the middle slit is removed by closing it (it is ½ wavelength out of phase), the top and bottom slit waves are added and intensity increases.

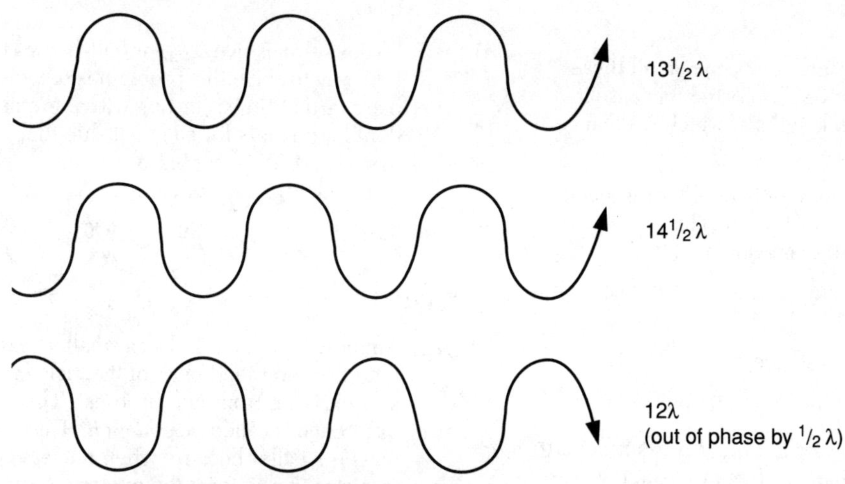

16. **B** The largest velocity corresponds to the largest angle away from normal. Light bends farther from normal in media of smaller "n" values.

17. **D** Since P and Q are 180° out of phase, they destructively interfere or cancel.

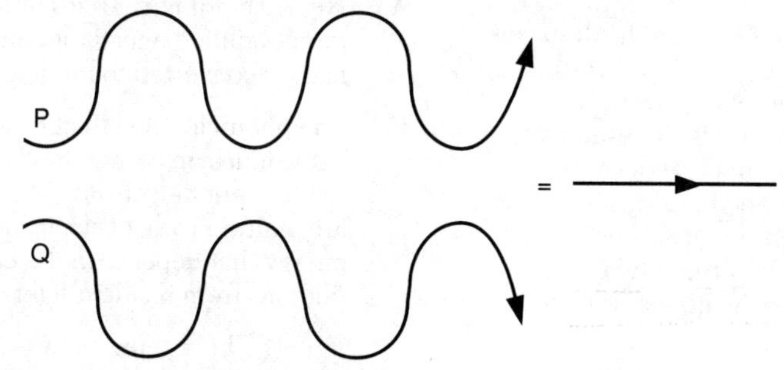

If P and Q were less than 180° out of phase, they would partially interfere.

18. **B** The frequency heard by a stationary observer behind the source is:

$f' = 1000[(10{,}000 + 0)/(10{,}000 + 2000)]$
$= 10{,}000/12$

The frequency heard by a stationary observer in front of the source is:

$f'' = 1000[(10{,}000 + 0)/(10{,}000 - 2000)]$
$= 10{,}000/8$

$f'/f'' = (10{,}000/12)/(10{,}000/8) = \frac{2}{3}$

19. **D** $\lambda = v/f = (300\text{ m/s})/(440\text{ s}^{-1}) = 0.75\text{ m}$.

20. **B** $T = 1/f \quad 0.2s = 1/f$
$f = 5\text{ Hz or }5\text{ s}^{-1}$

21. **A** The important equation to know for this problem is: $v = (F/\mu)^{1/2}$.

String one: $v = (F/\mu)^{1/2}$

String two: $v = (2F/2\mu)^{1/2} = (F/\mu)^{1/2}$

22. **A** For a simple harmonic oscillator, $y = A\sin(2\pi ft + \theta)$ or $y = A\sin(\omega t + \theta)$.

In this problem, $y = 3\sin(\pi t + 0)$.

Here then: $A = 3$ and $\omega = \pi$.

Since $\omega = 2\pi f = \pi$, $f = \frac{1}{2}$ and $T = 1/f$ or 2 s.

23. **C** Pendulum frequency and period depend only on the variable of pendulum length.

24. **A** $L = n\lambda/4$, where n = odd number. Because $\lambda = 10$ cm, try odd values for n and see at what lengths resonance occurs:

$n = 1 \quad L = (1)(10)/(4) = 2.5\text{ cm}$

$n = 3 \quad L = (3)(10)/(4) = 7.5\text{ cm}$

n = 5   L = (5)(10)/(4) = 12.5 cm.

n = 7, etc.

Note that resonance occurs at all these lengths. Choice A is correct because it includes two lengths at which resonance occurs.

25. **C** This simple harmonic oscillator is given by $y = A \sin(2\pi ft + \phi)$. This displacement is sinusoidal.

26. **B** Beat frequency = $|f_1 - f_2| = |50\text{ Hz} - 30\text{ Hz}| = 20\text{ Hz}$.

27. **B** Power $\approx 1/(\text{distance})^2$

Power $\approx 1/(10\text{m})^2 = 0.01\text{ m}^{-2}$

28. **A** Water, iron, and granite, for example, are all denser than air. Because sound waves move by compression, sound travels slowest in the least dense of the media given, that is, air.

29. **B** A β-particle: An α-particle:

$\beta = {}^{0}_{-1}e$ $\alpha = {}^{4}_{2}He$

β-decay involves a unit increase in atomic number whereas α-decay gives a two-unit decrease in atomic number. The net result of these two decays is a one-unit decrease in atomic number. β-decay gives no change in atomic weight; α-decay gives a four-unit decrease in atomic weight. The net result of these two decays is a four-unit decrease in atomic weight.

Y must be ${}^{36}_{17}Y$

Thus, ${}^{36}_{17}Y \longrightarrow {}^{32}_{15}I + {}^{4}_{2}He \longrightarrow {}^{32}_{16}X + {}^{0}_{-1}e$

30. **A** Gamma particles are photons with no mass.

31. **B** 30 days equals five ($t_{1/2}$) or half-lives. The best way to solve this problem is to work backward in time. Starting with 3 grams, double the mass for each half-life that has passed. Working backward, one arrives at 96 grams of element X.

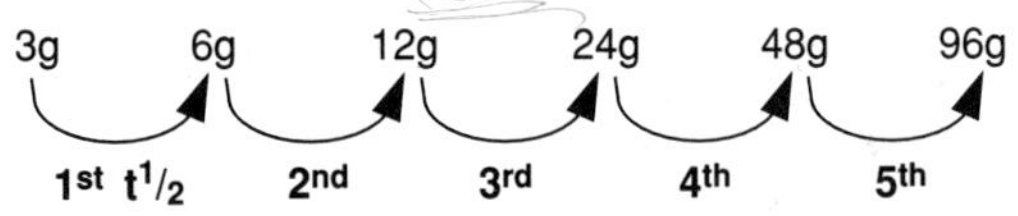

32. **A** An individual nucleus has a smaller mass than the combined mass of its nuclear particles (neutrons and protons). This difference is called *mass deficit.* This mass is smaller because when nuclear particles join to form the nucleus, a small amount of mass is converted to energy. This energy is called *binding energy.* It helps to hold the nucleus together.

33. **D** $(186.2\text{ MeV})(1\text{ amu}/931\text{ MeV})(1.66 \times 10^{-24}\text{ g}/1\text{ amu}) = 3.3 \times 10^{-25}\text{ g}$.

34. **A** Higher binding energies imply greater forces holding nuclei together. Nuclear mass is converted to binding energy.

35. **C** The photoelectric effect occurs when light shines on certain metals and causes them to emit electrons. These electrons are emitted with a velocity and kinetic energy that depend on the energy of photons from incident light:

$$hf = ½\, mv^2 + w$$

where: hf = energy in light photons, $½\, mv^2$ = KE of emitted electrons, and w = energy for electrons to break away from the metal.

HIGH-YIELD REVIEW QUESTIONS

Section III: Physics

Comprehensive Quiz

1. Light traveling through air strikes and enters water. How does the speed of light in air (S_a) compare with the speed of light in water (S_w)?

A. $S_a = S_w$
B. $S_a > S_w$
C. $S_a < S_w$
D. Cannot be determined

2. Light strikes an air–water interface and is reflected. How does θ_1 compare with θ_3? Assume that θ_1 = 42 degrees.

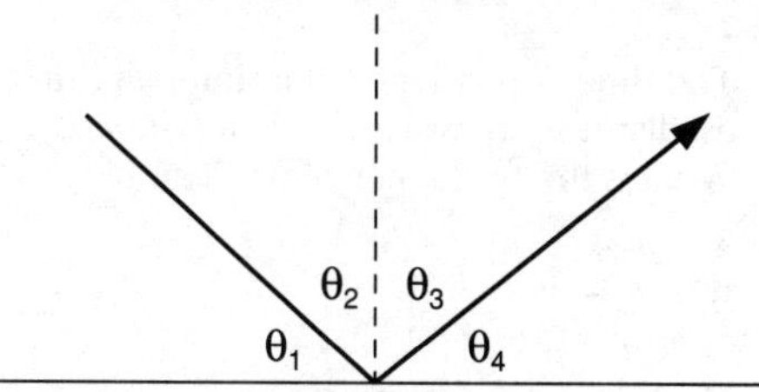

A. $\theta_1 = \theta_3$
B. $\theta_1 > \theta_3$
C. $\theta_1 < \theta_3$
D. No comparison

3. Light traveling in air strikes an interface with a glucose solution, n = 2.0, at an incidence angle of 45 degrees. The direction of the refracted wave is:

A. 45 degrees.
B. $\sin^{-1} \sqrt{2}$ degrees.
C. $\sin^{-1} \sqrt{2}/2$ degrees.
D. $\sin^{-1} \sqrt{2}/4$ degrees.

4. In question 3, did light bend? If so, why?

A. Yes; v decreased.
B. Yes; *f* decreased.
C. No; v increased.
D. No; *f* increased.

5. A man is standing in front of a plane mirror. His image is:

A. virtual, erect, m = 1.
B. virtual, inverted, m = ½.
C. real, erect, m = 1.
D. real, inverted, m = ½.

6. An object is placed 3 meters from a concave mirror that has a 2-m center of curvature. At what point is the image of this object found, and is it inverted or upright?

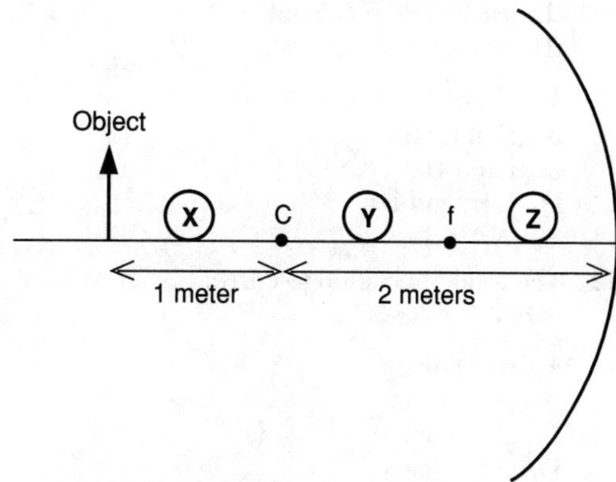

A. X, upright
B. X, inverted
C. Y, upright
D. Y, inverted

7. An object is placed between the focal point and lens in the following converging lens system. Where is the image?

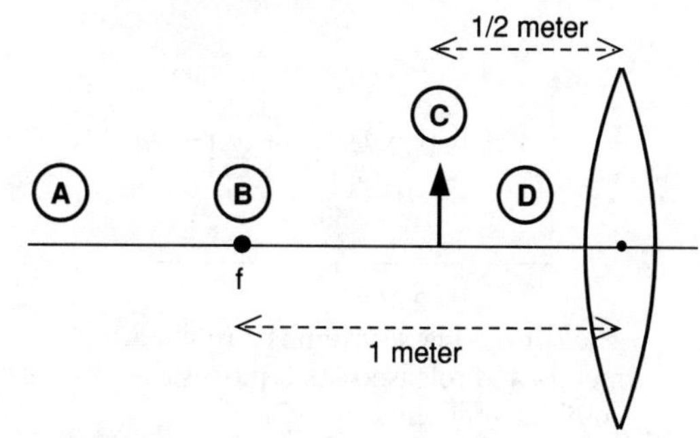

A. A
B. B
C. C
D. D

8. A lens of high power in diopters has:

A. long focal length.
B. long center of curvature.
C. short focal length.
D. short center of curvature.

9. In an alternating circuit, the maximum circuit current is 10 amps. What is the average or *rms* current?

   A. $5\sqrt{2}$ amps
   B. $10\sqrt{2}$ amps
   C. $2\sqrt{5}$ amps
   D. $2\sqrt{10}$ amps

10. Common forms of opposition to current in AC circuits are:

    I. resistance.
    II. capacitive reactance.
    III. inductive reactance.

    A. I and II
    B. II and III
    C. I and III
    D. I, II, and III

11. The total opposition to current in an AC circuit is called:

    A. resistance.
    B. inductance.
    C. impedance.
    D. capacitance.

12. An oscillating spring–mass system is shown in the following diagram. The mass glides along a horizontal frictionless surface. The spring reaches a maximum length of 4 meters. When completely relaxed, the spring is 2 meters long. The mass of the block is 10 kg, and $k = 5$ N/m.

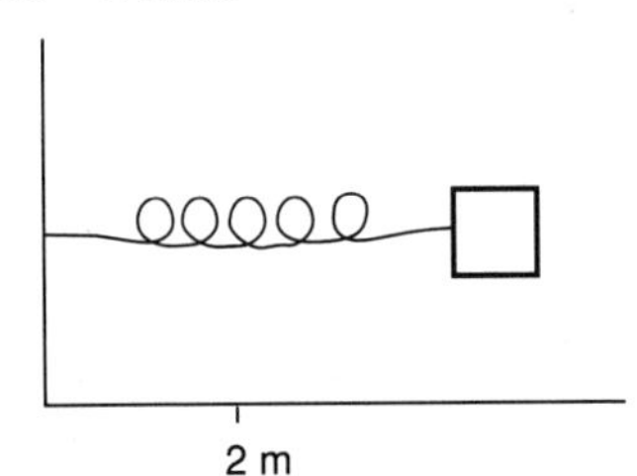

    The spring–mass system is stretched to 4 meters and released. As it passes the equilibrium point:

    A. PE > KE.
    B. KE > PE.
    C. KE = PE.
    D. cannot be determined

13. In the system described in question 12, what is the PE of the spring–mass system after it passes the equilibrium point 1 m?

    A. $\frac{5}{2}$ J
    B. 5 J
    C. 2 J
    D. 10 J

14. What force does the spring exert on the block at maximum stretch length?

    A. $^{-}20$ N
    B. $^{-}10$ N
    C. $^{+}20$ N
    D. $^{+}5$ N

15. What is the wavelength of a wave that has a period of 0.1 sec, and a velocity of 300 m/s?

    A. 20 m
    B. 30 m
    C. 40 m
    D. 50 m

16. Which can pass through a vacuum?

    A. Light
    B. Sound
    C. Both
    D. Neither

17. The time dependence of a simple harmonic oscillator is equivalent to $5\sin(10t + \phi)$. What is the frequency of oscillation?

    A. $5/\pi$
    B. $5/2\pi$
    C. 5π
    D. None of the above

18. Which factors are needed to calculate the period of a simple pendulum in oscillation?

    I. Mass of bob
    II. Length of pendulum
    III. g
    IV. Amplitude of swing

    A. I and II
    B. I and III
    C. II and III
    D. I, II, and III

19. A sound is produced under water. The sound then propagates to the surface of the water, and some of the sound is transmitted into the air. The velocity of sound in water is 1450 m/s, and the velocity in air is 330 m/s. What happens to the *frequency* and wavelength when the sound passes from water to air?

    A. Frequency and wavelength are constant.
    B. Frequency is constant; wavelength increases.
    C. Frequency is constant; wavelength decreases.
    D. Frequency increases but wavelength decreases.

20. Several key terms describing mechanical wave motion follow. Which term–definition is NOT matched?

A. Beats—Waves of different frequencies are superimposed to give slow, regular changes in the intensity of vibration.
B. Overtone—Any resonance frequency that is higher than the fundamental (lowest) resonance frequency of a system.
C. Harmonic—An overtone that is an integral multiple of the fundamental frequency.
D. None of the above

21. A sound wave of wavelength 0.2 m has a frequency of 5 cycles per second. What is the velocity of the sound?

A. 0.04 m/s
B. 10.0 m/s
C. 4.0 m/s
D. 1.0 m/s

22. A resonant system has a fundamental frequency of 100 Hz. If the next higher frequencies that give a resonance are 300 Hz and 500 Hz, the system can be:

A. a pipe open at both ends.
B. a pipe closed at both ends.
C. a string vibrating between two fixed points.
D. a pipe open at one end and closed at the other.

23. Two 80-dB horns play simultaneously. What is the level of sound that can be heard? (log 2 = 0.3)

A. Between 80 and 85 dB
B. Between 95 and 100 dB
C. Between 120 and 130 dB
D. Between 155 and 165 dB

24. If sound travels 1100 ft/s, the smallest resonance frequency of sound waves in a 6-inch tube closed at one end is:

A. 1100 Hz.
B. 2200 Hz.
C. 550 Hz.
D. 3300 Hz.

25. When two waves with the same wavelength and ½ wavelength out of phase travel in opposite directions in the same path, the result is:

A. echoes.
B. standing waves.
C. destructive interference.
D. beats.

26. What is the frequency of sound perceived by a stationary observer as a train approaches at 20 m/s blowing its whistle of 800-Hz frequency?

A. 750 Hz
B. 700 Hz
C. 900 Hz
D. 850 Hz

27. Radioactive substance X has a time constant magnitude of 0.1 sec^{-1}. If the mass of X at time zero is 128 grams, approximately how much X remains after 42 seconds?

A. 2 grams
B. 8 grams
C. 32 grams
D. 64 grams

28. When $^{238}_{92}U$ decays to $^{234}_{92}U$, it undergoes three decays with intermediates $^{234}_{91}Pa$ and $^{234}_{90}Th$, but not necessarily in that order. The sequence of decay is:

A. beta, gamma, alpha.
B. gamma, beta, beta.
C. alpha, beta, beta.
D. beta, alpha, beta.

SOLUTIONS

Section III: Comprehensive Quiz

1. **B** The light (wave front) slows upon entering a medium that has a greater index of refraction. Recall that n = 1 for air and n = 1.33 for water. The wave slows, and therefore bends toward the normal as shown in the following diagram:

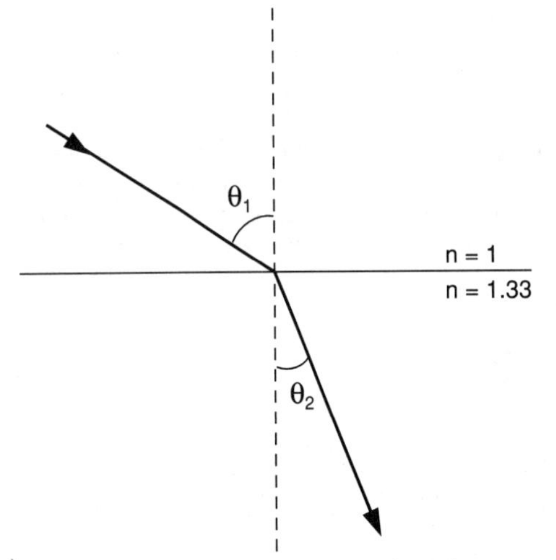

2. **C** Here, θ_1 = 42 degrees. Therefore, $\theta_2 = 90 - \theta_1$ = 48 degrees. Because the light is reflected, the law of reflection applies ($\theta_i = \theta_R$), and θ_3 is also 48 degrees. The best answer is choice C because 42 degrees (θ_1) is less than 48 degrees (θ_3).

3. **D** Snell's law tells us that $n_1 \sin\theta_1 = n_2 \sin\theta_2$. Recall that sin 45° = $\sqrt{2}/2$. The calculation becomes: $(1)(\sqrt{2}/2) = 2 \sin\theta_2$; $\sin\theta_2 = \sqrt{2}/4$; $\theta_2 = \arcsin \sqrt{2}/4$.

4. **A** Light bends toward the normal because the velocity of the wave front decreases upon entering the medium of higher n value.

5. **A** The image that one sees when looking into a plane mirror is upright (erect), has the same height (m = 1), and seems to be located "behind" the actual mirror (virtual). The following diagram shows the image created by a plane mirror.

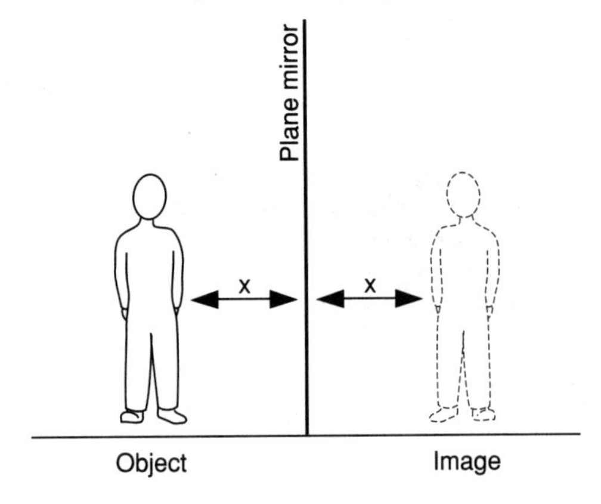

6. **D** This question can be solved using either the thin-lens equation or the "three-ray analysis technique." Both methods show the location of the image. Using the thin lens equation: $1/p + 1/q = 1/f$; $1/3 + 1/q = 1/1$; $q = 3/2$ m. The image is in front of the mirror. Using the "three-ray analysis technique," the following diagram can be drawn (note that the image is in a location also predicted by the thin-lens equation):

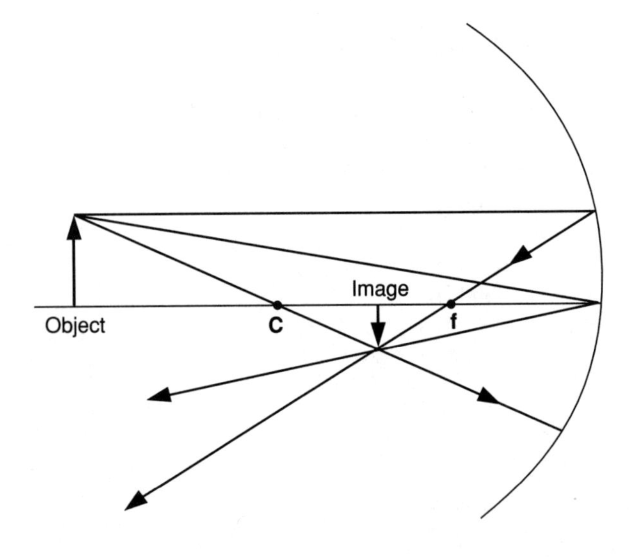

7. **B** Using the thin-lens formula, $1/p + 1/q = 1/f$, the result is $1/0.5 + 1/q = 1$. Therefore, $1 = {}^-1$ m. The negative sign means that a virtual image is produced (on the left side of the lens).

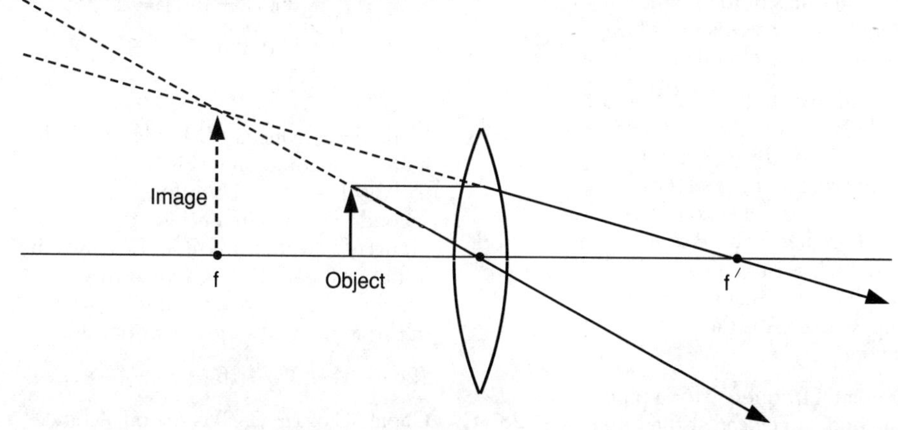

8. **C** Power = 1/f. Therefore, increasing P is observed with decreasing f. Note that power is measured in diopters. To use this equation, focal length (f) must be in meters.

9. **A** $I_{rms} = I_{max}/\sqrt{2} = 10\sqrt{2}/2 = 5\sqrt{2}$ amp.

10. **D** Resistance, capacitive reactance, and inductive reactance all oppose current in AC circuits.

11. **C** Impedance (Z) equals total opposition to current. Resistance, capacitive reactance, and inductive reactance are individual elements that resist current in AC circuits.

12. **B** When the spring is fully stretched and still at rest, the spring–mass system has all potential energy and no kinetic energy. Once released, potential energy of the spring–mass is converted into kinetic energy. As the mass passes the equilibrium point, (X_0 = 2m), the kinetic energy of the spring–mass is maximal. Therefore, KE > PE. As the spring–mass system passes the equilibrium point, the spring becomes compressed, the spring–mass system slows, and kinetic energy is converted back to potential energy.

13. **A** $PE = \frac{1}{2} kx^2 = \frac{1}{2}(5 \text{ N/m})(1 \text{ m})^2 = \frac{5}{2}$ Nm $= \frac{5}{2}$ J.

14. **B** $F = {}^-kx = {}^-(5 \text{ N/m})(2 \text{ m}) = {}^-10$ N.

15. **B** $v = f\lambda = \lambda/T$. Therefore, $\lambda = vT = (300 \text{ m/s})(0.1 \text{ s}) = 30$ m.

16. **A** Only light can pass through a vacuum. The transmission of light is mediated by photons. Transmission of sound requires molecule-to-molecule collision. A vacuum, by definition, lacks this needed matter.

17. **A** For simple harmonic motion, displacement about the equilibrium point equals $A \sin(2\pi ft + \phi)$. From the information given, $10 = 2\pi f$. Solving for f gives $f = \frac{10}{2}\pi = 5\pi$.

18. **C** $T = 2\pi\sqrt{(L/g)}$. Therefore, solving for the period of a simple pendulum in oscillation requires knowing both L (pendulum length) and g.

19. **C** Recall that sound requires molecule-to-molecule contact to propagate, and that water is more dense than air. The question says that sound travels faster under water than in air. Therefore, the velocity of the sound wave decreases as it moves into open air. Knowing that $v = f\lambda$, f is constant when a wave moves through different media, and v decreases, it logically follows that λ must also decrease.

20. **D** Choices A, B, and C are correct definitions. Therefore, choice D is the best choice.

21. **D** Because $v = f\lambda$, $v = (5 \text{ Hz})(0.2 \text{ m}) = 1$ m/s.

22. **D** Note that the fundamental frequency (f_o) equals 100 Hz. Also note the pattern of the resonant frequencies. The question tells us that the next higher frequency is 300 Hz, which means that the

fundamental frequency has increased by a factor of 3. The second harmonic, or second higher frequency over the f_o giving resonance, is five times the f_o. Note that the resonance frequencies are odd-integer multiples of the f_o. In addition, these pipe systems must have one open end and one closed end.

23. **A** Estimate the answer to this question. If 10 radios all play at 80 dB simultaneously, a 10-dB increase in intensity is heard, or a total sound intensity of 90 dB. If only two radios play, one can predict that the sound intensity will be less than 85 dB. Remember that this situation is a log relationship. Choice A is the only reasonable one.

24. **C** The fundamental frequency for a pipe open at one end and closed at the other is as follows: $f = nv/4L = (1)(1100\ m/s)/(4)(5\ m) = 2200/4 = 550\ sec^{-1} = 550\ Hz$.

25. **B** Choice B is the definition of standing waves. Recall that beats are waves of different frequencies that are superimposed to give slow, regular changes in the intensity of vibration. Destructive interference may occur when there is superposition of two harmonic waves or when waves are of the same period but out of phase with each other. Echoes involve wave reflection.

26. **D** As the train approaches the observer, the perceived frequency (f′) will be greater than the true frequency (f).

$$f' = f(1 + v_o/v)/(1 - v_s/v)$$
$$= 800(1 + 0)/(1 - 20/330)$$
$$= 800(330/310)$$
$$= 850\ Hz$$

27. **A** If the decay constant $(\lambda) = 1/10\ sec^{-1}$, the half-life of compound X is $t_{1/2} = 0.693/0.1 = 7$ sec. Therefore, in 42 seconds, six half-life decays occur. If 128 grams of X were present, only 2 g will be left after 6 half-lives. Arrive at this conclusion by drawing a simple diagram of the sequential series of six decays:

$128 \rightarrow 64 \rightarrow 32 \rightarrow 16 \rightarrow 8 \rightarrow 4 \rightarrow 2.$

28. **C** Choice C or (α, β, β) is the only choice that complements the particular scenario given. Remember that an alpha decay is a helium nucleus or $^{4}_{2}He$. The beta particle is an electron, or $^{0}_{-1}e$, and a gamma particle is a photon (massless).

$$^{238}_{92}U \rightarrow\ ^{234}_{90}Th + \boxed{^{4}_{2}He} \rightarrow$$

$$^{234}_{91}Pa + \boxed{^{0}_{-1}e} \rightarrow\ ^{234}_{92}U + \boxed{^{0}_{-1}e}$$